Methuen's Monographs on Biological Subjects

General Editor: G. R. de Beer, M.A., D.Sc.

PLANT VIRUSES

METHUEN'S MONOGRAPHS ON BIOLOGICAL SUBJECTS

F'cap 8vo., 3s. 6d. net each

General Editor: G. R. de Beer, M.A., D.Sc.
Fellow of Merton College, Oxford

SOCIAL BEHAVIOUR IN INSECTS. By A. D. Imms, M.A., D.Sc., F.R.S.
MICROBES AND ULTRAMICROBES. By A. D. Gardner, M.A., D.M., F.R.C.S.
MENDELISM AND EVOLUTION. By E. B. Ford, M.A., B.Sc.
THE BIOCHEMISTRY OF MUSCLE. By D. M. Needham, M.A., Ph.D. (5*s.* net).
RESPIRATION IN PLANTS. By W. Stiles, M.A., Sc.D., F.R.S., and W. Leach, D.Sc., Ph.D.
SEX DETERMINATION. By F. A. E. Crew, M.D., D.Sc., Ph.D.
THE SENSES OF INSECTS. By H. Eltringham, M.A., D.Sc., F.R.S.
PLANT ECOLOGY. By W. Leach, D.Sc., Ph.D.
CYTOLOGICAL TECHNIQUE. By J. R. Baker, M.A., D.Phil.
MIMICRY AND ITS GENETIC ASPECTS. By G. D. Hale Carpenter, M.B.E., D.M., and E. B. Ford, M.A., B.Sc.
THE ECOLOGY OF ANIMALS. By Charles Elton, M.A.
CELLULAR RESPIRATION. By N. U. Meldrum, M.A., Ph.D.
PLANT CHIMAERAS AND GRAFT HYBRIDS. By W. Neilson Jones.
INSECT PHYSIOLOGY. By V. B. Wigglesworth, M.A., M.D.
TISSUE CULTURE. By E. N. Willmer, M.A.
PLANT VIRUSES. By Kenneth M. Smith, D.Sc., Ph.D.

In Preparation

BIRD MIGRATION. By W. B. Alexander, M.A.
MYCORRHIZA. By J. Ramsbottom, O.B.E.

Other volumes to follow

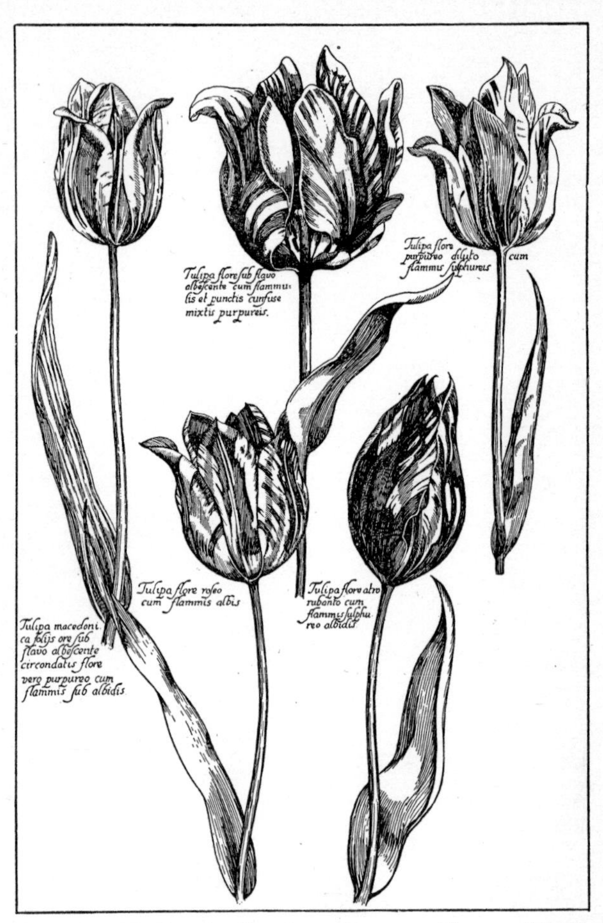

Mosaic or 'Broken' tulips from *Theatrum Florae*, originally published in 1622, and recently identified by Savage as the work of the painter Daniel Rabel. 'Tulipa macedonica', at the left, suggests the possibility of 'broken' foliage, both in the drawing and in the descriptive polynomial. (*Redrawn from* McKay and Warner [48]).

PLANT VIRUSES

BY

KENNETH M. SMITH, D.Sc., Ph.D.

POTATO VIRUS RESEARCH STATION, SCHOOL OF AGRICULTURE, UNIVERSITY OF CAMBRIDGE

WITH ELEVEN ILLUSTRATIONS

METHUEN & CO. LTD.
36 ESSEX STREET W.C.
LONDON

First published in 1935

PRINTED IN GREAT BRITAIN

PREFACE

IN this small book the aim of the author is to bring to the notice of workers in other branches of science, and particularly botanists and entomologists, some of the more interesting and important facts of the study of plant viruses. In their study lie problems of fundamental biological importance, such as their affinities with enzymes, their intimate relationship with the living cell, their curious and interesting association with insects, and finally the possibility that they may represent a new and simple form of life. It is by detailed attention to such problems as these that virus workers hope one day to find the answer to the question—what is a virus?

The references at the end of the book are not intended to be exhaustive but are selected as being representative of the main lines of plant virus research.

The writer wishes to express his indebtedness to Mr. Brooks, F.R.S., Dr. Salaman, F.R.S., Mr. Bald and Mr. Bawden for their advice on certain chapters and to Mr. Doncaster for making the drawings for the illustrations.

K. M. S.

Potato Virus Research Station,
School of Agriculture,
Cambridge

CONTENTS

CHAPTER I

INTRODUCTORY

Historical

REFERENCES to disorders of plants, which are now known to be due to virus infection, occur in the literature of very early times. The first record is a description published in 1576 by Charles l'Ecluse or Carolus Clusius [22] of a variegation in the colour of tulips which is now called 'breaking' and is recognized to be due to an aphis-transmitted virus. 'Broken' tulips are figured in *Theatrum Florae* published in 1622; these illustrations have been identified as the work of the painter Daniel Rabel [48] and one of his pictures is reproduced in the frontispiece. On the left of this illustration may be seen the striping of a leaf which is a characteristic symptom of the disease. A somewhat later account published in *Traité des Tulipes* about 1670 contains the first suggestion that the variegation in the flower colour might be due to a disease. About 100 years later the agricultural community is found to be seriously perturbed about a disorder, or rather group of disorders, in the potato crop known collectively as the 'curl disease' which is only now being slowly dissociated into its constituent virus entities. The potato 'curl' seems to have made its first appearance in this country round about the year 1770 and various attempts to explain its nature and propound a cure were made (letters to the Agricultural Society of Manchester, 1792). In 1802 the suggestion was first made that insects might in some way be connected with potato curl. This suggestion was later destined to be amply confirmed and the relationship of plant viruses with insects is now a fundamental part of the study of these interesting disease agents.[72]

About 1868 the variegated plant *Abutilon* (Malvaceae) appeared in Europe and became popular as an ornamental plant. By grafting scions of variegated individuals to green shoots of normal plants it was discovered that this variegation was infectious. Now the variegation in *Abutilon* is known to be due to a virus infection.

In 1886 Mayer described a disease of the tobacco plant which he called *Mosaikkrankheit* and this term is now widely used for describing the mottling type of virus disease. Mayer showed that this disease could be inoculated mechanically to a healthy tobacco plant. Two years later Erwin F. Smith proved that the disease of peach yellows was infectious and could be transmitted by budding.

It was not, however, till 1892 that a scientific demonstration of the existence of a filterable virus was made. Iwanowsky working with the mosaic disease of tobacco, described by Mayer, proved that the sap from such a diseased plant was capable of inducing the mosaic disease in healthy tobacco plants after it had been passed through a bacteria-proof filter candle. This discovery, however, passed comparatively unnoticed and the work was repeated seven years later by Beijerinck who then propounded his theory of a *contagium vivum fluidum*.

In the year 1901 the first definite proof that some connexion existed between insects and plant viruses was obtained by Takami, a Japanese worker, who succeeded in transmitting the ‘ stunt ’ disease of rice from diseased to healthy plants by means of an insect, the leaf-hopper *Nephotettix apicalis*.

Potato mosaic was first seen by an American, Orton

by name, in Germany in 1911, the irony of this being, as Murphy has pointed out, that not a single commercial American potato plant has since been found which is free from mosaic. To Orton in his paper published in 1914 is owed the specific name of leaf-roll and the first mention of potato mosaic and streak.

Organized and intensive study of the general plant virus problem, however, is mainly a post-war development and most of it has been inaugurated during the last ten years.

As regards the important question of potato virus diseases, modern developments may be said to commence with the work of Quanjer in 1916 who showed that potato leaf-curl, now known as leaf-roll, was a communicable disease. This was followed in 1920 by the discovery of Oortwijn Botjes that leaf-roll was transmitted by insects though the actual insect responsible was not identified till later.[71]

For many years it was considered that the degeneration or ' running out ' of potato varieties after a period of vegetative reproduction was due to ' in-breeding ' or ' senile decay ', and it was the discovery that potato leaf-roll was an infectious disease which gave the death-blow to this theory and paved the way to the true explanation that the degeneration of potato stocks is due to contamination with viruses.

In 1923 the insect-proof glasshouse equipment for the study of potato virus diseases was completed at Wageningen in Holland and this equipment has been copied and modified in many countries since that date.

Characteristics of Plant Virus Diseases

There exists no acid test by which a disease in a plant can be immediately recognized as due to virus infection. There are, however, certain features which may be said to be characteristic of virus diseases. As a rule the symptoms are uniformly spread over the plant including the flowers and sometimes the fruit and are most pronounced on the young leaves and near the growing points.

In the majority of virus diseases the infective agent is systemic in all parts of the affected plant with the usual exception of the seed, though this is by no means an invariable rule. All plant viruses are infectious and can be transmitted from diseased to healthy plants by several methods. In some cases the only successful method of transmission is by grafting or budding.

Virus diseases are more common in herbaceous than in woody plants, although peach yellows and mosaic, plum pox, buckskin of cherries and one or two other virus diseases of woody plants are known.

Symptoms are very variable and take the form of mottling, streaks and lesions on the leaves or on the stems, distortion of the whole plant and alterations in the colours of the flowers. There may also be reduction in the vegetative parts used for reproduction.

The absence of disease symptoms in the roots of affected plants is probably due to the lack of chlorophyll in these tissues. The pathological picture which is ordinarily associated with plant virus diseases, i.e. the streaks, mottlings and distortions mentioned above, may be referred to the destructive effect of the virus upon the chlorophyll apparatus or to disturbances in metabolism resulting from injury to that apparatus.[35, 90]

It is unusual for plants to be killed outright by virus diseases unless infection occurs in the young or seedling stage. Viruses do, however, appear to pave the way to attack by other disease agents in certain cases. This is shown by the increased susceptibility of virus-diseased potatoes to infection by 'blight' (*Phytophthora infestans*) as compared with virus-free potatoes of the same variety. Mosaic-diseased sugar-cane seems also to be less resistant to infection with various disease organisms than healthy cane.

The Chief Plant Virus Problems in Different Countries

The chief plant virus problems in various countries are of great interest and significance. In the British Isles the most important group of viruses from the economic point of view is that affecting the potato which, in England, causes the deterioration of potato stocks after one or two seasons' growth and necessitates the import of fresh 'seed' from Scotland. This question is further dealt with in Chapter IX.

The potato 'curl disease' of the eighteenth century is now known to consist of two groups of virus diseases; the first, called leaf-roll, is sporadic in its appearance but is a serious disease from the farmer's point of view when it does appear. Its chief insect vector is the aphis *Myzus persicae* Sulz. The second or mosaic group is a difficult problem to elucidate as the various component viruses have not yet all been isolated. Four or five however have now been separated out and symbols allotted to them. These mosaic viruses appear to be of two kinds, the X and Y types [73]: the latter are transmitted by the aphis *Myzus persicae*, but there is still some doubt as to the exact method of spread of the X type viruses. It is possible that a species of thrips is the vector. In the field a potato plant is rarely affected with a single mosaic virus, for example the familiar disease known as 'crinkle' is a complex of the X and Y types of virus.

The tomato is liable to infection with several viruses of the mosaic type which, however, are quite distinct from those attacking the potato. The best known are tomato mosaic, tomato streak or stripe, tomato yellow or aucuba mosaic and tomato spotted wilt. The last-named was only discovered in England in 1929 but it has already become one of the worst troubles the tomato grower has to deal with, and the ability of this virus to attack all manner of ornamental plants renders it of considerable importance to the grower with 'mixed' glasshouses. The insect vector of this virus is the thrips, *Thrips tabaci*.

The stocks of certain strawberry varieties, notably Stirling Castle and Royal Sovereign, appear to be approaching a state of degeneration only comparable to the degeneration of potatoes already mentioned. While progressive virus infection may not be the only reason for this it is probably one of the chief causes.[16]

Other important virus diseases in the British Isles are those affecting raspberries, peas, lettuces, cucumbers and hops.

In France the chief virus problem appears to be that of the potato crop, particularly leaf-roll which is widespread in south-western and south-eastern France. Tobacco crops in the Gironde are also heavily infected with viruses of the mosaic and ring-spot types and in some districts mosaic of the vine is of importance.

In Germany, in addition to potato viruses and others common to European crops generally, the sugar-beet is infected with a disease known as leaf-crinkle or *Krauselkrankheit*, caused by a virus similar to but not identical with the notorious curly-top of sugar-beet in the U.S.A. The insect vector of the German virus is a plant bug, *Zosmenus quadratus*.

Probably the most important virus disease in the U.S.A. is the curly-top of sugar-beet mentioned above. This was first recognized as a malady of major economic importance in 1899, in California. Since that time frequent and very destructive outbreaks have occurred in most of the sugar-beet areas west of the Rocky Mountains. In the regions where it occurs as epidemics curly-top is the most destructive of all beet diseases. The insect vector of the virus is the beet leaf-hopper, *Eutettix tenellus* Baker, and the geographical distribution of the disease is therefore the same as that of this insect.[12]

Potato viruses also present a serious problem in the U.S.A.; all the European potato viruses appear to occur there together with others not found in Europe.[51] So prevalent in America is the potato mosaic virus—called in this country virus X—that it is known in the States as the 'healthy potato virus', being latent without visible symptoms in most of the American potato stocks.

Heavy losses in the head lettuce crop are caused by the virus of tomato spotted wilt in certain coastal districts in California. Infected plants are unmarketable and die prematurely.[84]

The mosaic disease of cucumbers and allied plants occurs commonly in the United States and Canada and has ruined the cucumber-growing industry in some areas. It was first reported in Ohio in 1902 and in Massachusetts in 1909, and since those dates has become widely distributed on Cucurbitaceae throughout the United States.

Two important virus diseases of another type occur in North America and Canada, these are aster yellows and peach yellows. The former virus affects the China aster (*Callistephus chinensis* Nees) and was first described about 1902. The aster is grown extensively in other parts of the world besides the U.S.A., especially in Europe and the Orient, but the aster yellows virus appears, so far, to be confined to North America and Canada. The insect vector is a leaf-hopper, *Cicadula sexnotata* Fall.

Peach yellows, also transmitted by a leaf-hopper, causes annually the loss of many peach trees in the north-eastern parts of the United States, while another virus also attacking the peach appears to be even more serious. This virus, which produces 'phony disease', is said to have caused the loss of more than a million peach trees in Georgia and to have spread right across the south-east in little more than a decade.

Mosaic of sugar-cane at one time caused great losses

to the sugar-growing industry, but much of this loss is now avoidable by the substitution of mosaic-resistant canes for susceptible varieties.

A disease of sandal, having all the characteristics of virus infection, is widespread in parts of southern India. 'Spike disease', as it is called, was first reported in Coorg in 1899 and has since spread in an eastward direction.

Perhaps the most startling sudden invasion of an economic crop by a virus disease is the recent development of 'leaf curl' in the cotton crops of the Sudan. This virus is transmitted by a species of whitefly (Aleyrodes), *Bemisia gossypiperda*. The first diseased plant was collected in the Gezira in the Sudan in 1923–4. In 1926–7 the spread of the disease continued in the Gezira. By the end of 1927–8, nearly 50 per cent of the plants were infected in areas where cotton had been grown for some years. The next two seasons saw the completion of the process of penetration and practically every cotton plant over the whole Gezira area, some 200,000 acres of cotton, showed more or less severe symptoms. It appears that Egypt is still free of the disease at all events so far as cotton is concerned.[1]

Virus diseases are not very prevalent in Egypt, the chief infection being mosaic of sugar-cane; bunchy-top of bananas occurs occasionally.

South Africa has its virus problems in mosaic of sugar-cane, transmitted by an aphis, *Aphis maidis*, streak of maize, transmitted by a leaf-hopper, *Cicadulina mbila*, and a virus disease of the necrotic type affecting tobacco which is thrips-transmitted and is probably due to the same virus as that causing spotted wilt of tomatoes. In this connexion it may be mentioned that pineapples in Hawaii are subject to a virus disease, called yellow spot, which is disseminated by a thrips. This disease may also be caused by the spotted wilt virus.

Spotted wilt is a serious problem in Australia where tomatoes constitute an important field crop. First recorded near Melbourne by Brittlebank in 1915, the disease has now spread all over Australia as well as the U.S.A. and Europe where it causes heavy losses to the tomato-growing industry.

Bunchy-top of bananas is also a disease of some importance in Australia. The insect vector is the dark banana aphis *Pentalonia nigronervosa*.

From this short survey may be seen the world-wide distribution and great economic importance of the virus diseases which affect economic crops of all kinds from bananas to tulips.

Increasing Spread

That plant virus diseases appear to be more prevalent and of greater economic importance than even a decade ago cannot be denied. It is true that the greater intensive study of the subject during the past ten years has made the recognition of a virus disease more certain but there must be other factors as well. The increasing facilities for trade and transport together with the search for new varieties of plants favour the dissemination of viruses. Vegetative parts of plants such as the tubers of dahlias or potatoes can harbour one or more viruses without indication of their presence, and these offer a ready means of transfer from one country to another. Previous to 1929 spotted wilt of tomatoes was unknown in England, now it has been recorded from practically every county. This virus could easily have been introduced in dahlia tubers or on some other ornamental plant from which it could be disseminated by the thrips already present in the country. Similarly with the German beet disease, *Krauselkrankheit*, the potential insect vector, a Tingid bug, is already present in England and it only needs the two to be brought into contact to achieve the spread of this beet disease.

Increasing cultivation of the land can be instrumental in aiding the spread of viruses by the destruction of weeds and other wild plants which may be the natural hosts both of a virus and its insect vector. The insect is thereby driven to change its habits and find a new food plant, at the same time bringing with it the virus infection. This fact may account for the sudden appearance of leaf-curl in the cotton crops of the Sudan. Where a virus disease has once become established, however, destruction of susceptible weeds and the wild host plants of insect vectors is a valuable means of combating the disease.

Viruses with a wide host range such as those of curly-top of sugar-beet, aster yellows and tomato spotted wilt are gradually penetrating to new host plants in which they produce what is to all intents and purposes a new disease. Thus the virus of tomato spotted wilt is known to cause diseases of different aspect in dahlias, cinerarias, asters, arum lilies, amaryllis, broad beans, cow peas, and lettuces, in addition to the disease of the tomato with which this virus is chiefly associated.

Finally it is possible that viruses may become adapted to new plants and to new insect vectors and thus increase their host range and multiply their means of dissemination.

Economic Importance

A few examples will suffice to show the great losses caused annually by virus diseases in important crops. The poor quality generally of English 'seed' potatoes is largely due to their contamination with viruses and the acknowledged superiority of Scotch 'seed' to their comparative freedom from these disease agents. It is not, perhaps, too great an exaggeration to say that the immense trade in Scotch 'seed' potatoes is based upon the scarcity or absence of a certain species of greenfly (*Myzus persicae* Sulz.) in the 'seed'-growing areas of Scotland. One can also predict that a few more hot and dry summers like those of 1933 and 1934 will see a movement of this virus-bearing aphis up into the more northerly 'seed'-producing districts of Scotland with a consequent deterioration in the quality of the potato crops in those areas.

Fig. 1 shows the average yields of sugar-beet in tons

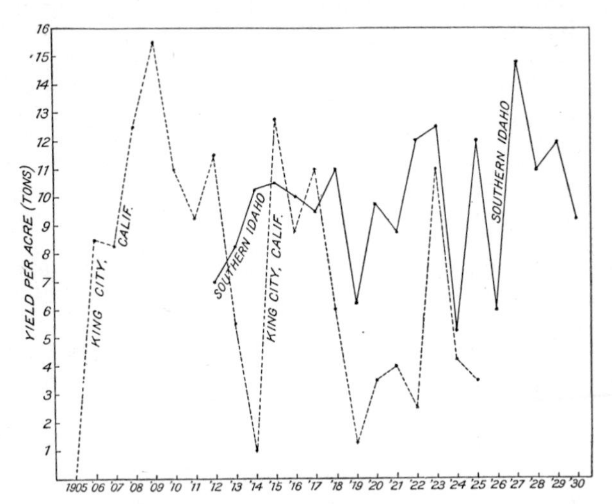

FIG. 1.—Average yields of sugar-beet in tons per acre for 1905–30 in the King City district of California and in southern Idaho. The serious decreases in yields were caused by curly-top injury. (*After* Carsner)

per acre for the years 1905–30 in the King City district of California and in southern Idaho, U.S.A. The serious decreases in yield were due to curly-top injury. In the California district in the years 1909, 1912, 1915 and 1917, which were relatively free from curly-top, the average yields were 15·6, 11·5, 13·0 and 11·3 tons per acre respectively, in contrast with the disastrous curly-top years 1914 and 1919 when the yields were 1·0 and 1·4 tons per acre respectively.[12]

The ‘ phony disease ’ of the peach began to assume an alarming aspect in the orchards of Georgia in 1915. By 1927, 99 per cent of the trees were affected and in 1928 some 600 acres of trees were cut down and grain was planted in their place. During 1931 and 1932 the inspectors of the Phony Peach Eradication Campaign found and cut down 3,980 diseased trees.[36]

The export trade in Easter Lilies from Bermuda owes its decline to infection with a virus disease of the stunting or rosetting type. The trade declined from 13,803 boxes of bulbs exported in 1896 to 2,357 boxes exported in 1913. By a partial elimination of the disease the export trade increased to 6,000 boxes in 1927, but the European market was almost entirely lost.[55]

Spike disease of sandal in India is believed to cause an annual loss of five to six lakhs of rupees, while comparable losses are caused by the mosaic disease of cucumbers in the U.S.A. and by leaf-curl of cotton in the Sudan.

CHAPTER II

GENERAL TECHNIQUE OF PLANT VIRUS STUDY

Glasshouse Technique

SINCE the majority of plant viruses are disseminated from diseased to healthy plants by insects, it is essential that any glasshouse intended for the study of the subject should be adequately insect-proof. It is, however, quite possible to overestimate the importance of the insect-proofing, at all events in temperate countries, and to reduce the light and air circulation below the minimum required for successful cultivation of the plants by excessive quantities of wire gauze over windows and ventilators. It must be remembered that whatever studies are being made of the virus the final test has to be made on the plant itself since the reaction of the host is almost the only criterion of the presence of the virus. For this test therefore healthy and actively growing young plants are necessary.

There are two main types of glasshouse at present in use for virus studies. In one type there is a central corridor with the insect-proof chambers on either side (Fig. 2). These chambers should not be less than 10 feet by 9 feet, otherwise heat and moisture become excessive owing to the reduction of the air circulation through the wire gauze. In the other type of glasshouse the chambers are placed back to back without a central corridor. Each chamber opens into a small porch about 6 feet by 2½ feet by 3 feet, which itself has a door opening to the exterior. A person entering the porch closes the outer glazed door before opening the inner door, the panels of which are replaced by wire gauze. This procedure is intended to preclude the entry of insects either by flight or on the clothes of the person entering. In actual practice, however, it has been found that in the height of summer when insects are abundant it is usually necessary to keep the outer porch doors

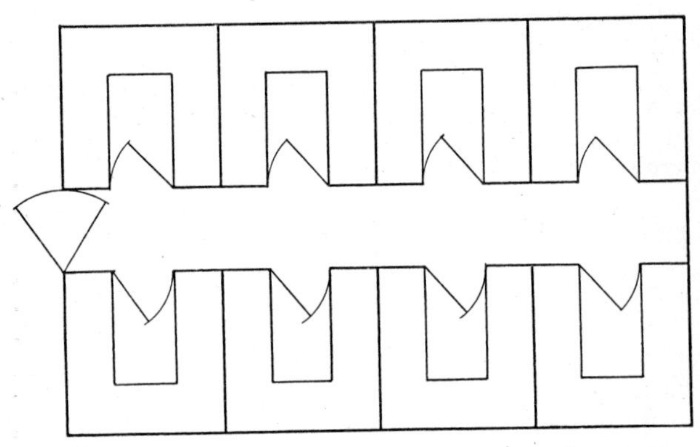

FIG. 2.—Diagram of glasshouse suitable for plant virus study: the insect-proof chambers open out of a central corridor

open in order to reduce the temperature inside the chambers. This, of course, removes the *raison d'être* of the porches. A further disadvantage is the difficulty of entering the porches with trays of pot plants. After eight years' experience of these porches the writer has formed the conclusion that they are redundant and that the better type of house is the one with the central corridor. The criticism has been made that this corridor is liable to form a trap for insects which, having once entered, are likely to escape into the chambers. Efficient routine fumigation, however, should prevent this.

In the British Isles glasshouses are most liable to infestation with aphides, whiteflies, thrips and red spider, and these may gain access in spite of insect-proofing. Routine fumigation is therefore necessary and should be regularly practised. In the Cambridge glasshouses a nicotine fumigant is used against aphides and thrips and two proprietary insecticides against whiteflies and red spider.

The most usual method of proofing a glasshouse against insect invasion is shown in Fig. 3. The ventilators in the roof open into a shallow trough lined with phosphor-bronze gauze (No. 40 mesh), while the side ventilators should be covered in the manner shown in the diagram. Where there is direct entrance into the glasshouse it is a good plan to have, behind the glazed door, a second light frame door of phosphor-bronze gauze. This allows the outer door to be kept open during hot weather.

It is always advisable to use sterilized soil, particularly when dealing with the more infectious sap-transmissible viruses. There are several types of soil sterilizers which are suitable ; the one at present in use at Cambridge is of the box type and is electrically heated.

Experimental Transmission

A plant virus may enter its host through an extremely trivial wound, the breaking of a trichome is sufficient to allow the virus of tobacco mosaic to infect a plant, but it appears to be an accepted fact that a wound of some sort is essential to permit the entry of a virus. While this is the view generally held, it should be mentioned that some work recently published in America [21] describes the successful transmission of the virus of tobacco mosaic merely by spraying a suspension on to healthy tobacco plants.

This seems to show that the virus is able to enter by the stomata and so infect the plant.

Apart from the above-mentioned possibility there

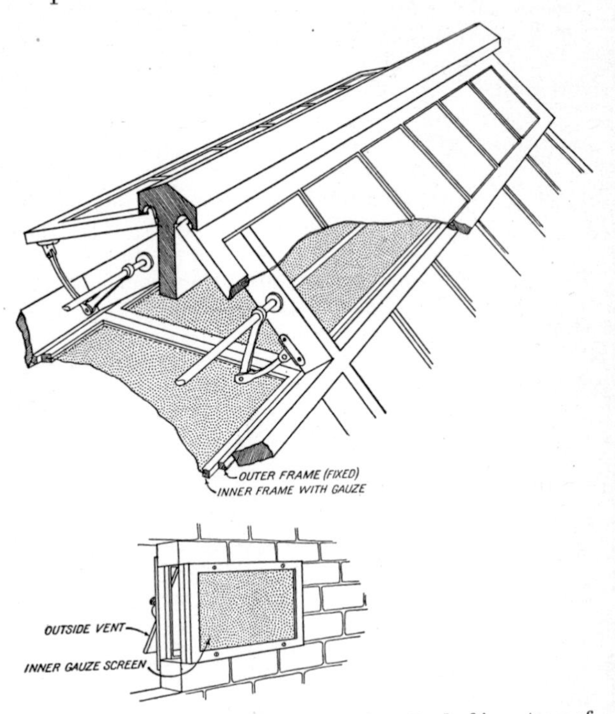

Fig. 3.—Sketch illustrating the usual method of insect-proofing a glasshouse for plant virus study

are three methods of virus transmission, by grafting, by inoculation, and by the agency of insect vectors, and the usual technique employed in these methods of transfer will be briefly described.

By Grafting. In order to make a successful graft

it is essential that actual organic union between stock and scion be effected. All plant viruses can be transmitted by grafting, though it does not necessarily follow that a graft-transmissible disease is due to a virus.

There are several methods of grafting, varying according to the type of plant used. For plants with soft sappy stems like potatoes, the best method is the ordinary cleft graft in which the stem of the scion, cut to a wedge shape, is inserted in a cleft in the stock. The graft is then bound with fine rubber tape and the loose end secured with a drop of ordinary rubber solution. Rubber tape is more suitable than bast because it does not constrict the stem of the plant but gradually perishes and falls away by the time organic union is completed. For other types of plants such as strawberries or lilies, inarching may be employed. This method consists essentially in removing by means of a razor-blade a small slice from the side of each runner or stem, as the case may be, and then binding them together as before, leaving the roots of each plant undisturbed.

In the study of spike disease of sandal where a woody plant is the subject, other methods have been devised. In ' patch-grafting ' a piece of diseased bark, with or without a bud, is inserted in a space cut in the bark of a healthy tree and the whole bound firmly with raffia. Another method of transmitting the virus of spike, not strictly grafting, consists of the insertion of a piece of infected leaf tissue under the bark of the healthy stock.[77] This method of virus transmission, however, is said to succeed only when the cut midrib of the leaf forms organic connexion with the tissues of the cut bark. Spike disease can also be transmitted by a ' ring bark graft ' in which the twigs of the tree to be inoculated are ringed by removing the bark all round to a width of three-quarters to one inch. Bark of corresponding dimensions from an infected tree is substituted and the whole bandaged.[60]

There are two types of tuber grafting which may be practised with potatoes, tulips and similar plants. ‘ Core-grafting ’ consists in the removal, by means of a cork-borer, of a core from the infected tuber and its insertion in a hole made in the healthy tuber with a cork-borer one size smaller. A somewhat similar method can be employed for transmitting a virus disease of the sugar-cane. Alternatively the cut surfaces of the two halves of diseased and healthy tubers respectively may be placed in contact and bound together with raffia. In the case of potato tubers the diseased half should be disbudded.

By Inoculation. The word inoculation is generally used to describe the introduction of virus sap into the tissues of a healthy plant. It has also, however, a wider meaning embracing any method of disease transmission. In this short description it is used in its former sense. Not all plant viruses are sap-transmissible, those that are belong mostly to the mosaic group and they vary somewhat in their response to different methods of inoculation. Scratching through a drop of virus sap into the leaf of the plant to be infected is the most obvious method and this is by no means the most efficacious. Many mosaic viruses can be transmitted by this method but the actual scratching plays a small part in the transmission. The less drastic the inoculation the more efficient, within limits, is the virus transmission, for it must be remembered that a virus requires a living cell for its multiplication, and therefore the less mutilation there is the better.

Tests have been carried out to discover whether scratching through a drop of mosaic extract was as effective as rubbing.[32] Twenty-five separate leaves were scratched, ten times each, through drops of the fluid. On these leaves the same virus was rubbed into similar small areas. Great numbers of lesions appeared from the rubbing but only 13 lesions were found on the 250 needle scratches. Scratching through mosaic extract appears slightly more effective than dropping virus into scratches already made, but neither method is as efficient compared with light rubbing. The virus does not easily enter wounds made before its application but requires wounds made in its presence, the entry into such wounds being apparently instantaneous. That the entry of virus into wounds is almost instantaneous is shown by the fact that washing the surface of a leaf immediately after inoculation in no way affects the subsequent development of the disease.

Various implements can be used for inoculating plants; the pestle, with which the infected leaves used as a source of inoculum have been crushed, serves the purpose very well if used lightly. Small pieces of muslin, cotton-wool or the cut edges of filter-paper dipped in virus sap and rubbed gently over the surface of healthy leaves, give a high percentage of positive infections. Spatulae with a ground glass surface or small pieces of rubber sponge may also be used. It is a good plan to support the leaf to be inoculated on a piece of waxed paper or a wooden label to avoid possible contamination by the fingers. A very gentle rubbing achieves the best results with some viruses, and the addition of an abrasive like fine sand or carborundum powder facilitates transmission to certain host plants. In most experimental transmissions with viruses it is important to use young and vigorous plants.

Culturing Insect Vectors

Because of the close relationship between insects and plant viruses a fundamental part of plant virus study is concerned with this relationship and some knowledge of the best methods of handling insects is necessary. Roughly speaking, there are two classes of insect vectors known at the present time: the first class contains the very minute thrips and the second class consists of various species of *Hemiptera* or sucking plant-bugs such as aphides, leaf-hoppers and whiteflies. In keeping stock cultures of an insect it is a good general rule that whenever possible the insect should be fed on a plant which is immune to the virus under study. With thrips and aphides this is easily accomplished since both are omnivorous feeders and have a wide host range. At Cambridge, for example, it has long been the custom to culture the aphides used in potato virus studies upon cabbage and irises, both of which are immune to potato viruses though both, it may be mentioned, are susceptible to other aphis-borne viruses.

The actual insect-cages used vary with the type of insect. Cellophane is a useful material for covering the cages which may have either wooden or metal frames. For infecting sprouting potato tubers or young seedlings by means of aphides, a suitable cage can be made with a glass chimney of the stable-lamp type covered at one end with cellophane or fine muslin. The flange at either end gives support when the chimney is pressed into the soil surrounding the plant. This type of cage may also be used for experiments with thrips unless it is necessary for the thrips to be kept under observation during pupation; in this case a 'Micro-cage' may be made as follows. A piece of thick felt about $\frac{3}{4}$ inch square with a $\frac{3}{8}$-inch diameter hole in the centre is laid upon the leaf of the experimental plant and the thrips are placed inside the ring. A square of stiff cellophane is placed below the leaf and a similar piece over the hole in the felt; the cage is then completed by the adjustment of a strong stationery clip.[57]

For the transfer of the insects themselves a moistened camel-hair brush is used for the handling of aphides and thrips while for the more active leaf-hoppers and whiteflies a ‘ catcher ’ is required. This consists of a piece of glass tubing covered at one end with muslin and fitting closely into a second piece which is drawn out pipette-like at the top. A piece of rubber tubing is attached to the end of the inner glass tube. By means of this apparatus leaf-hoppers and whiteflies may be drawn by suction into the first tube and lightly blown out again as required.

It has been stated earlier that not all plant viruses are sap-inoculable, indeed quite a number are only transmissible by the agency of specific insects, grafting excluded. In order to study the *in vitro* properties of such viruses it is necessary for subsequent transmission that their insect vectors should imbibe them in solution outside their plant hosts. In its simplest form this technique consists in feeding the insect upon the virus suspension, to which some sugar solution has been added, contained in a thin-walled sac made of animal mesentery or similar substance. Various modifications have been devised in which the membrane covers the end of a wide tube or inverted filter funnel containing the virus fluid and the insects are then induced to suck through the membrane and imbibe the fluid much as if they were feeding upon a leaf. Some workers have substituted a thin section of paraffin wax or the peeled-off epidermis of iris leaves for the animal mesentery. By means of this technique the physical properties of certain non-inoculable viruses such as those of curly-top of sugar-beet and streak of maize have been studied.

Separation of Viruses in a Disease Complex

Virus diseases of plants are sometimes caused not by one virus alone but by two or more co-existing in the same host, and failure to recognize this fact is liable to cause the greatest confusion when such a disease complex is being studied. Virus complexes occur most frequently in the potato mosaic group and it is largely this fact which makes the group such a difficult one to study. Such mixtures may be detected and the component viruses isolated by several methods, firstly by taking advantage of possible differences in certain physical properties such as resistance to ageing in extracted sap, filterability, thermal death-point, &c. Another method is to take advantage of possible differences in host range whereby inoculation to a series of differential hosts may eliminate one of the component viruses. A third means of separation is by the agency of the insect vector, as it does not necessarily follow that the same insect is the vector of both or all the viruses composing a disease complex. Indeed, it was the selective power of virus transmission possessed by an insect vector which first gave the clue to the composite nature of certain potato mosaic diseases.[73]

One or two examples may help to make the matter clearer. Two potato mosaic viruses, designated X and Y, frequently occur in combination, producing in the potato a severe distorting disease of the 'crinkle' type. If the chief insect vector of potato viruses, the aphis *Myzus persicae*, be fed upon a potato plant so affected and then transferred to a healthy plant, symptoms due to virus Y alone develop on the latter, but if a sap-inoculation be made from diseased to healthy plant, then the disease complex due to both viruses develops, since the two viruses are sap-transmissible. Further work upon X and Y viruses has shown that the two may be separated in the following ways: (1) Keeping the expressed sap in a test tube at room temperature for 48 hours eliminates the Y virus, which only remains viable in extracted sap under these conditions for 24–30 hours.

(2) By filtration through specially prepared collodion membranes the Y virus may be held back.

(3) Heating the expressed sap to 60° C. destroys the

Y virus, as its thermal death-point is lower than that of X.

(4) By the power of selective transmission possessed by the aphis *Myzus persicae*, the Y virus may be obtained alone.

(5) Inoculation of the virus complex to *Datura Stramonium* filters out the Y virus because this plant is immune to infection by Y.

(6) Inoculation of the virus complex to certain varieties of the garden Petunia eliminates X, as this plant is resistant, though not immune, to X.

The foregoing is rather an extreme case and so many methods of separation were only found to be possible after a careful investigation of the properties and behaviour of the two viruses.

Two other viruses occurring together in the tobacco plant afford another case in point. One of these is tobacco mosaic and the other an undescribed 'ringspot' virus, so called because it produces rings with a central spot on the host plant. Separation of these two was effected by inoculation to a plant which would eliminate one of the pair; two types of plant were used. The cowpea (*Vigna sinensis*) while immune to tobacco mosaic allowed the ringspot virus to become systemic whence it was re-inoculated to tobacco. Similarly *Nicotiana glutinosa* although susceptible to the virus of tobacco mosaic localizes it on the inoculated leaves and allows no systemic invasion, but the ringspot virus passes to the young leaves, where it can be obtained free of the other virus. The ringspot virus can be eliminated from the mixture by filtration and possibly, after further investigation, by other methods.

CHAPTER III

NATURAL MODES OF TRANSMISSION

ALL plant virus diseases are infectious and can be transmitted from diseased to healthy plants in a variety of ways, some viruses more easily than others. The artificial methods of transmission by grafting and sap-inoculation have already been discussed in the previous chapter so that only the natural means of dissemination will be dealt with here.

By Seed. Seed transmission, curiously enough, is a comparatively rare occurrence and only five or six authentic cases are recorded, of which mosaic of bean and certain other leguminous plants like clover are the best known. Seed from mosaic beans, however, do not always give rise to diseased plants, the percentage varying from 13 to 50.

The reasons for the comparative rarity of virus transmission by seed are not known, but it has been suggested that in the case of tobacco mosaic the virus may become inactivated by adsorption to the seed storage proteins.[19] In bean mosaic some seeds fail to transmit the virus while others from the same pod produce infected seedlings. A possible explanation of this irregularity may be found in the vascular anatomy that characterizes the bean pod. Starting from the apex of the pod, the successive ovules arise alternately from the opposite edges of the carpel, resulting in a bilateral arrangement of the seeds in the pod. This alternate attachment of the ovules to the opposite edges of the carpel might possibly explain the occurrence of infected and non-infected bean seeds within the same pod. The suggestion has also been made that uneven distribution of the virus in the bean plant might account for this irregular seed transmission.[53]

This latter suggestion can hardly be applicable to many mosaic diseases affecting Solanaceous plants as it is a commonplace that the virus is always present in the fruits and on the seed coat, but such seeds seem to produce healthy plants. Presumably the virus does not reach the embryo, possibly because of its anatomical isolation in the seed. Nevertheless, infection might sometimes arise through the cotyledon becoming contaminated on rupturing the seed coat and this may be the explanation of the occasional instances of apparent seed transmission of tomato and tobacco mosaic which do seem to occur. In curly-top of sugar-beet, however, the virus appears to be confined to the phloem elements and to be unable to flourish in the parenchymatous tissue. This fact may account for its non-transmission by the seed since there is no vascular connexion between the mother plant and the young sporophyte.[6]

Another anomaly is the transmission of a virus by seed of certain plant species only. The virus of a tobacco ringspot appears to be transmissible only by seed of petunia, although the virus has a very wide host range [30]; similarly cucumber mosaic virus is said to be transmitted by the seed of the wild cucumber, *Echinocystis lobata*, and of the muskmelon, but not through the seed of the cultivated cucumber.[18, 40] In the infectious variegation of *Abutilon* spp. seed transmission has been recorded but it seems to be a rare occurrence. There is a complete absence of the variegating virus in the seeds of *Abutilon regnelli* but a low percentage of infected seedlings has been obtained from other varieties of *Abutilon*.[41]

Transmission by the agency of pollen is said to occur in the case of a disease of *Datura* [9] and also in mosaic of bean,[61] but further confirmation is needed of this phenomenon.

TRANSMISSION THROUGH THE SOIL

A good deal of doubt still exists as to whether viruses can be transmitted through the soil, although positive statements to this effect have appeared from time to time in the literature.

It is questionable whether viruses ever remain viable for long periods in the soil away from their host plant. Probably the virus of tobacco mosaic is capable of retaining its infective power in pieces of tissue in the soil and healthy plants might become contaminated by contact with these. Such transmission, however, would necessitate an injury of some kind to the healthy plant by some other agent. Soil-inhabiting insects are the most probable agents by which soil transmission could be effected and these by their nature and habits are unlikely to be efficient vectors. It is of course possible that an insect like the wireworm in its peregrination below the soil-level might bring infective material on its jaws to a healthy plant and so induce disease, but, as will be seen in a moment, the part played by biting insects like wireworms in plant virus transmission is exceedingly small. At the same time some such possibility cannot be altogether excluded.

Wheat seedlings are said to become infected with mosaic through the roots, but so far there is no evidence that a vector of an animal nature is concerned with this spread.[88]

On the whole, soil transmission probably plays a very small part in the general spread of plant viruses.

Transmission through Vegetative Reproduction of Infected Plants

Since the majority of plant viruses are systemic in their hosts, that is all organs of the plant with the usual exception of the seed are invaded, the virus persists from year to year in the organs of vegetative reproduction such as tubers, rhizomes and bulbs.

There are many examples of such propagation of virus diseases. The outstanding case of course occurs in the potato plant, the tubers of which become infected year by year with viruses till a complete state of ' degeneration ' has set in. It is this propagation of viruses by the tuber which necessitates replacement by Scotch ' seed ' after one or two years' growth in England. All tuberous, bulbous or rhizomatous plants behave in a similar manner. Dahlias infected with spotted wilt or mosaic, irises and daffodils infected with breaking or ' stripe ', reproduce the disease year after year.

Propagation by cuttings or suckers from infected plants also results in the production of diseased plants. This occurs with mosaic of raspberry and bunchy-top of bananas.

Transmission by Insects

Plant viruses depend very largely upon insects for their transmission in the field and the relationship between insect and virus is one of considerable interest and scientific importance. This relationship is discussed in more detail in Chapter VI, so that for the present attention is confined to a survey of the type of insect most concerned with, and to some of the factors governing, the spread of viruses in the field.

If the various types of insects implicated in virus transmission be examined it will be found that this power appears to be rather a specialized one and is possessed for the most part by insects which feed in a particular way. These insects belong to the plant-sucking group, Hemiptera, and it is a subdivision of this group, Hemiptera-Homoptera, which contains the majority of the insect vectors. There is one other important insect vector, the thrips, which belongs to a slightly different order, the Thysanoptera. This insect is also of the sap-feeding type but does not possess the long suctorial proboscis found in the Hemiptera (see Fig. 8).

Exclusive of the Thysanoptera practically the whole of the insect vectors are contained in the order Hemiptera and can be classified as follows : *Hemiptera-Heteroptera. Tingidae*, Lace bugs. In this family there is one species, *Zosmenus quadratus* Fieb., which is the vector in Germany of leaf-curl or leaf-crinkle of sugar-beet. *Hemiptera-Homoptera. Jassidae* (*Cicadellidae*), Leaf-hoppers. The leaf-hoppers are slender active insects usually tapering posteriorly, they are as a group efficient vectors of plant viruses in countries other than the British Isles. Eight species of leaf-hoppers have been definitely identified as the vectors of as many different viruses, there are also other records which cannot yet be regarded as authentic. Such important virus diseases as curly-top of sugar-beet, aster yellows, peach yellows and streak of maize are transmitted by leaf-hoppers.

In the *Fulgoridae* or lantern flies, one species, *Peregrinus maidis*, is the vector of corn stripe. Next come the *Aleyrodidae* or whiteflies, of which the tomato whitefly is the best-known example in this country. Whiteflies are very small moth-like insects in which the wings are whitish or clouded and the body and wings more or less mealy. One or more species of this group, mostly *Bemisia* spp., have been shown to be the vectors of cotton leaf-curl, tobacco leaf-curl and a mosaic disease of cassava. Lastly come the *Aphididae*, the greenflies; this family of insects is more concerned with virus transmission than any of the other groups. Aphides have been identified with the spread of about twenty-five different plant viruses and of these twenty-five, one species, *Myzus persicae* Sulz., is capable of spreading fourteen. Aphis-transmitted viruses are mostly of the 'mosaic' type, though there are exceptions. The mosaic diseases of crucifers, of sugar-beet, of beans, of celery, of dahlias, of tulips, of sugar-cane and of raspberry are all aphis-transmitted viruses, as are also potato leaf-roll, yellow dwarf of onions and potatoes and bunchy-top of bananas, diseases of a slightly different kind.

There seems indeed to be some correlation between the type of insect and the kind of virus disease it transmits. Thus in addition to the 'mosaic' group of viruses largely transmitted by aphides, there is the 'yellows' group disseminated by leaf-hoppers; a type of disease causing malformation such as 'curling' of which whiteflies are the vectors and possibly a group of ringspot viruses carried by thrips. How far such a classification of viruses by their insect vectors is a genuine correlation and how far it is merely fortuitous cannot at the moment be determined.

In the British Isles the only insect types so far identified as vectors of plant viruses are aphides and thrips.

Factors governing Spread in the Field

The spread of a virus through a crop is not merely a simple question of the propinquity of insect vector and virus disease. Many other factors are involved, the chief of which are the climatic and environmental conditions pertaining at the time.

It is proposed here to discuss these factors and illustrate their reactions on the spread of virus diseases by four representative types of insect vectors—aphides, thrips, leaf-hoppers and whiteflies. Aphides, particularly *Myzus persicae*, are the vectors of several virus diseases of the potato and it is in relation to this crop that their movements in the field are considered. The initial factors are the number of winged (alate) aphides which arrive from without the crop and the degree of mobility within the crop. Any conditions tending to increase or restrict movement of these vectors within the crop are of importance. Another point is the date of maximum infestation of the aphides in relation to maturity of foliage, important because of the difficulty of infecting fully-grown plants with virus diseases. In other words, the presence of a heavy infestation of aphis vectors in mid-August or later does not necessarily involve a high percentage of virus infection.[91]

If a potato crop is originally healthy the multiplication within the crop of wingless (apterous) aphis vectors would not be followed by the spread of virus; the important points therefore are the arrival of winged aphides, possibly already infected, from outside sources, and their subsequent movement about the crop. The movement of these alate aphides is governed largely by weather conditions, flight being greater in hot, dry weather and very much reduced under conditions of high humidity.

The chief aphis vector of potato viruses (*Myzus persicae*) hibernates on *Brassicas*, particularly savoys, and the presence of these plants near a potato crop is therefore detrimental to the health of the latter. In practice, however, areas in which a large quantity of *Brassicas* are grown are usually those in which potatoes also are largely grown.[50]

It will thus be realized that the presence of alternative host plants for insect vectors in the vicinity of susceptible crops plays a large part in the infection of those crops with virus diseases.

With a thrips-transmitted virus like spotted wilt of tomatoes, a somewhat similar condition of affairs exists, the arrival of infected thrips from outside sources being the important factor. If infection spreads from a local disease centre already in the crop, diseased plants will tend to be concentrated into groups; but if, on the other hand, the position of a plant does not affect its chances of infection, diseased plants will be found distributed at random when infection arrives from outside the plot. Tests on field plots of tomatoes in Australia have shown that in some cases infection is more concentrated at the sides of a field near diseased and infected plants of other host species; in others the initial centres of infection were distributed at random, but, as the plants matured, infection tended to spread more to adjacent than to distant plants. Observations have also shown that in actively growing plants of a susceptible variety a positive relation exists between high daily incidence of the disease and high temperature twelve days earlier and between low incidence and low temperatures twelve days earlier. This relation suggests a greater effectiveness, or in other words, greater activity of the insect vector on hot days and a fairly constant incubation period of about twelve days of the virus in the plant. It has been established for some species of thrips that adults emerge from the pupal stage mainly on warm days and that there are minimal temperatures above which they fly and migrate. The most favourable temperatures for the dispersal of the thrips vectors of spotted wilt are not less than 75° F., and may be over 80° F. Wind probably plays a considerable part in long-distance dispersal on hot days. On the other hand, low temperatures delay the development of immature stages and reduce the activity of the adult thrips which are the principal vectors owing to the short range of movement of the larvae. These observations also emphasize the importance of isolating a tomato crop, so far as practicable, from overwintering sources of infection.[2]

In the case of a virus disease of tobacco in Java known as Kroepoek (leaf-curl) in which the insect vector is a whitefly (*Aleyrodidae*), it has also been shown that the chief source of virus infection for the insect lies outside the plot. This result has been arrived at by a comparison of the rates of spread between tobacco mosaic which is almost wholly transmitted by mechanical contamination by the labourer and tobacco leaf-curl where the sole vector is the whitefly. It has been shown mathematically that infection by the labourer as transmitter of mosaic rapidly increases in a cumulative manner during the growing period while infection of the Aleyrodids remains much more constant. The course of the mosaic disease is explained by supposing that infection of the labourer depends primarily on the number of mosaic-diseased tobacco plants present in the plot and that infection of labourers outside the plots is practically nil. On the other hand, the fairly marked constancy of disease increase of Kroepoek indicates that the source of infection for the *Aleyrodidae* lies mainly outside the plot and that compared with this, Kroepoek-diseased plants in the plot are of minor importance. As a control measure, therefore, the removal of Kroepoek-diseased plants from the plots will assist very little. The sources of infection *outside* the tobacco plots must be determined and destroyed as far as possible before the planting time for tobacco. That there is another factor in the movement of insect-borne viruses is shown by the great accumulation of the Kroepoek disease in the neighbourhood of villages and buildings. This appears to be due to the habit of the insect of collecting in protected situations.[83]

An unfavourable host-plant condition plays a part in governing the migration of an insect vector and consequently the spread of a virus disease. It has been found [13] that the movements of the beet leaf-hopper, *Eutettix tenellus*, which transmits the virus of curly-top, are largely governed by host-plant relationships. Two conditions are met with, although no sharp lines of demarcation between the two can be made. If conditions were favourable for hibernation of the insect the previous autumn, an extremely dry spring will prove disastrous to the beet growers. Under such conditions the alternate host plants such as mustard, *Salsola pestifer* and *Atriplex rosea* are killed off or do not germinate. On these occasions there is a general movement of *E. tenellus* from environments which are no longer suitable. The movements into beet fields which occur to some extent every year are more difficult to explain, but the same explanation, i.e. an unfavourable host-plant condition, probably holds true. There may, however, be other reasons closely connected with this, such as competition from other insects and an instinct to migrate, on the part of the spring brood, which may be a response to a stimulus brought about by maturing plant tissue.

It is interesting to note that the economic problem of curly-top in parts of Idaho, U.S.A., has been brought about by the introduction of certain host plants which have proved favourable for the development of large populations of *E. tenellus*. The years of high wheat prices stimulated the breaking up and cultivation of large areas of semi-arid land close to the irrigated tract. With the collapse of high prices these lands proved to be sub-marginal and were abandoned. Water shortage led to the abandonment of large acreages of land which were originally cleared of sagebrush with the intention of irrigating them. On most of these abandoned lands the prevailing plants are hosts for *E. tenellus*. Among these hosts the most important are *Salsola pestifer*, the Russian thistle, and mustard. The presence or absence of *S. pestifer* seems to determine in large measure the distribution and size of the population of *E. tenellus* entering hibernation, while mustard is an important spring host since it is from this plant that the migration takes place.

Wind plays a small part in the distribution of the leaf-hoppers unless they are disturbed, but severe winters reduce the numbers hibernating and in consequence are followed by good yields of sugar-beet in southern Idaho.[13]

From a consideration of the foregoing it will be seen that one important factor in the infection of a crop is the arrival of insect vectors from outside sources. Various factors govern this movement in which wind, high temperature, shelter and host-plant relationships all play a part. Distribution of the virus within the crop depends upon the degree of infection already present, the type of insect vector and the degree of maturity of the foliage at the time of maximum infestation.

Ecological studies have suggested the possibility of forecasting future epidemics of insect vectors, and this in time may prove of value in combating virus diseases.

Although insects are generally held responsible for the natural dissemination of plant viruses, there are some viruses which appear to spread without the aid of these agents. A good example is tobacco necrosis, a disease recently discovered and described.[75] The virus causing this disease appears spontaneously in insect-proof glasshouses and attacks seedlings and young plants of tobacco and *Nicotiana glutinosa*. Infection appears to spread in the complete absence of insect vectors and even when the plants are grown in sterilized sand or soil. The obvious explanation would of course be transmission through the seed were it not for the fact that the virus appears not to become systemic in its host and so is unlikely to reach the seed. It has been noticed that only such plants as tobacco and *N. glutinosa* whose lower leaves come into contact with the soil surface become *naturally* infected; plants such as tomato, *Datura* and cow-pea, in which the lower leaves are some distance off the soil, never become naturally infected although they are equally susceptible. This of course suggests that infection is conveyed by the soil except that it occurs equally in sterilized and unsterilized soil. The behaviour of such a virus appears at first sight to transcend the natural laws governing the spread of most plant diseases, although further investigation will probably throw light on much that is at present obscure. At the same time these facts, coupled with the extremely small size of the virus, 20 mμ, and its ability to withstand 48 hours in 99 per cent alcohol, lead not only to speculation on new modes of transmission but to doubts on the homogeneity of the plant virus group.

Recent work in America [21] has shown that healthy tobacco plants can be infected with tobacco virus 1 merely by spraying a suspension on to the plants with an atomizer. The writer has carried out similar experiments with the above tobacco necrosis and potato virus X and has obtained the same result. In the case of potato virus X the sprayed tobacco seedlings developed such large numbers of local lesions as to suggest that the virus was entering freely by the stomata. It is by no means improbable that such experiments may throw some light on the spread of certain viruses for which no insect vector can be found.

CHAPTER IV

THE VIRUS IN THE HOST

Effect on the Plant

THE diseases induced by viruses in their host plants are very varied in appearance, but most of them have in common an even distribution of the symptoms on the plant. This is due to systemic invasion of the tissues by the virus and is a point of difference from diseases caused by other infective agents in which symptoms are usually more localized.

Symptoms may be used, in spite of their variability, for a loose classification of virus diseases into the mosaic group, the 'yellows' group and a third miscellaneous collection of which the chief characteristic is overgrowth or abnormal stimulation of some part of the plant.

The name mosaic was originally applied by Mayer to the classical tobacco disease because of a fancied mosaic pattern of different shades of green and yellow on the leaves, and this name has been adopted to describe that type of virus disease which induces a chlorotic mottling. The mottling, however, is by no means the only symptom of 'mosaic' viruses, many of which have a destructive effect on the host cells, giving rise to a lethal necrosis. The alternative symptoms are dependent for their development upon various factors such as the species of the host plant, environmental conditions and degree of virulence of the virus. They also depend on the tissues in which they multiply, e.g. viruses exert an action, either inhibitory or destructive, upon the chlorophyll, which gives rise to the mosaic mottlings mentioned above. They are, however, capable of multiplying in plant roots or in etiolated stems kept in darkness, in other words, under conditions where chlorophyll is absent,[35] and under these conditions there are no symptoms.

In addition to chlorotic mottling and necrosis certain mosaic viruses have a third type of manifestation in the development, mostly on the leaves, of numerous concentric rings (Fig. 4) and wavy lines and patterns. With some viruses there may be as many as eight concentric rings with a central spot; they are usually necrotic, though chlorotic rings also occur. The majority of mosaic viruses seem capable of ring formation under some conditions but little is known of the underlying causes.

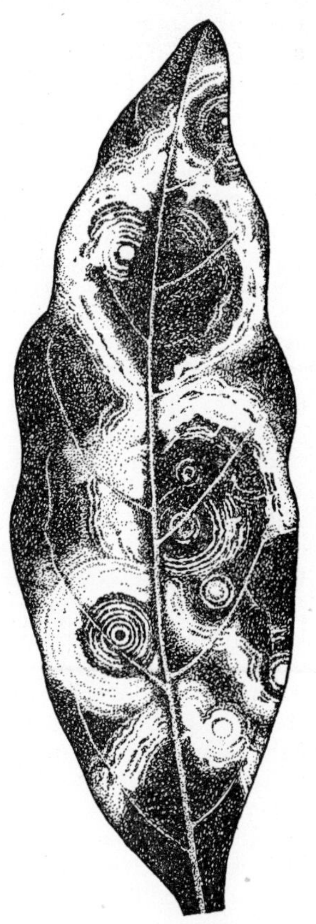

FIG. 4.—Leaf of a plant of *Solanum capsicastrum* infected with the virus of tomato spotted wilt, showing the concentric rings and line patterns characteristic of this virus on certain hosts

In aster 'yellows' there is no mottling, but chlorosis is general throughout the affected parts of the plant. The leaves show a tendency to stand upright instead of lying flat and are also frequently deformed ; the petioles are longer and the leaf blades narrower than in the normal plant.

In addition to the virus diseases causing mottling or yellowing of the leaves there is the third miscellaneous group mentioned above in which the affected plant is distorted or stimulated to abnormal growth. In potato leaf-roll the leaves are heavily charged with starch and in consequence they are stiff and leathery and roll inwards. Dwarfing and rosetting of the plant are common results of infection with some of these virus diseases, the internodal spaces of the stems are shortened and the leaves are brought close together. Such diseases are known as curly-dwarf, bunchy-top, stunt disease, *court-noué* and so forth.

In other cases the internodes are elongated instead of shortened so that a weak spindly stem is produced or numerous weakly stems may arise instead of one or two ; such diseases are known as 'witches' brooms'. 'False-blossom' of cranberries is a good example of this type of disease as it combines several kinds of malformation. As a rule the flowers show the symptoms most clearly and are usually rendered sterile. The calyx lobes of diseased flowers become enlarged, the petals are short and streaked with red and green and the stamens and pistils are abnormal. When the disease is severe the entire flower may be replaced by successive whorls of leaves or by a short branch. The leaves also show symptoms of the disease. Axillary buds which are usually latent are stimulated to produce numerous negatively geotropic branches with many crowded leaves giving rise to a witch's broom.[17]

Other examples of abnormal growth are given by 'giant hill' of potatoes which results in unusually large plants, by the galls occurring in the phloem of sugar-cane caused by 'Fiji disease' and by the production of elongated tubers in the 'spindle-tuber' disease of potatoes.

Some plants though systemically infected with a virus yet show no symptoms and appear quite normal. Such plants are known as 'carriers', and this phenomenon occurs commonly in certain potato varieties affected with particular viruses. If a scion from one such infected carrier be grafted to a healthy potato plant of another 'non-carrying' variety, the virus latent in the scion passes down into the stock, which develops visible symptoms. A potato variety which carries one mosaic virus may, however, produce normal symptoms when infected with another mosaic virus. The hop plant also affords an example of a carrier and certain varieties are frequently infected with a mosaic virus which produces no visible effect.

It will be realized that infected carrier plants are a source of danger in the field to other susceptible varieties.

In addition to the external symptoms described already certain pathological changes occur in the tissues of virus-affected plants. The most characteristic of these changes is the appearance of an inclusion, or so-called X-body, in the cells of plants affected by certain viruses. Intracellular inclusions were at first considered by a few workers to be the causative organism of virus diseases and were so described in America some years ago in connexion with bean mosaic.[52] The opinion is now generally held that X-bodies are an effect rather than a cause and should be looked upon as a reaction of the cell cytoplasm to the virus. The general appearance of these X-bodies is shown in Fig. 5. They are usually vacuolate, somewhat amoeba-like in shape and often bear a superficial resemblance to a protozoan. Further reference to these cell inclusions will be found in Chapter X, where they are compared with somewhat similar bodies associated with animal virus diseases.

Reference has been made in a previous paragraph to the effect of viruses upon chlorophyll; this effect is considered [69] to be an inhibition of plastid formation, but destruction of mature plastids also seems to occur.[74] Some viruses have a destructive action on the cells, especially the sieve tubes and companion

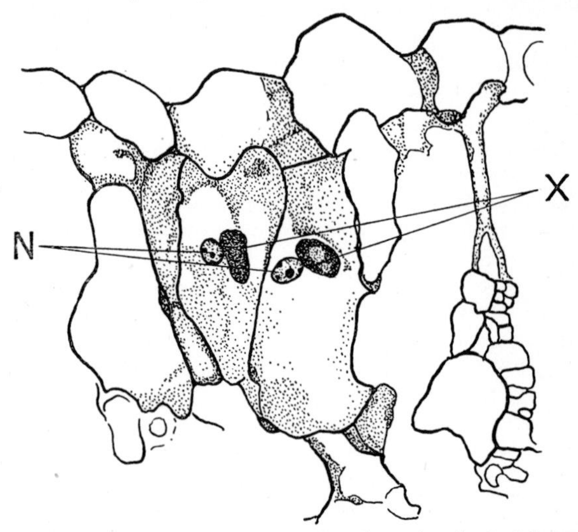

Fig. 5.—Part of a transverse section through a leaf of *Datura* infected with potato virus **X.** Note the two intracellular inclusions or **X**-bodies (**X**) close to the nuclei (**N**). (Highly magnified)

cells, and the phloem necrosis characteristic of potato leaf-roll is an example of this.

Local Lesions

Local necrotic lesions are produced by some viruses at the site of inoculation in certain hosts. This is particularly true of the tobacco mosaic viruses and some species of *Nicotiana*, especially *N. glutinosa*. In these plants the virus does not become systemic but remains localized at the point of inoculation. The use of local lesions allows the recognition of large numbers of successful transmissions on single plants. This method, which has been compared with Koch's plate method with bacterial cultures, makes possible comparative estimates of virus concentrations. When highly diluted samples of virus are used and very small numbers of lesions develop on the leaves of the host plant, it seems possible that each infection has resulted from a single virus particle.[31] At higher concentrations of the virus there is no direct and simple relationship between the concentration and the numbers of lesions produced, but within certain limits it is possible to tell which of two samples of virus is the more concentrated, and to gain some idea of their relative virus content. Important points to be observed in using the local lesion method for quantitative work are the adoption of a standard method of inoculation and the comparison of virus samples by inoculation on opposite halves of the same leaves or on single leaves arranged in such a way as to eliminate the extreme effects of variation in susceptibility.[66, 93] The kind and degree of this variation were shown by the statistical analysis of experiments in which plants of *Nicotiana glutinosa* were inoculated with tobacco mosaic virus and the numbers of lesions produced were counted. The data were submitted to reduction by the analysis of variance.[93] Plants differed greatly in their reaction to inoculation and a gradient of susceptibility was established between the different leaf positions. The nature of the gradient varied with different sets of plants. It was shown that the right and left halves of a leaf responded equally to the inoculation procedure used in the experiments.

It will be understood that the local lesion method of study, by its nature, is applicable only to those viruses which produce lesions at the site of inoculation.

Movement of the Virus in the Host Plant

The path followed by the virus in its passage through the plant is a question of great interest and importance and it has been studied by a number of workers. As regards the method of movement of a representative virus, like that of tobacco mosaic, there are two main views, which may be stated as follows:

(1) A progressive advance from the point of inoculation through the tissues of the plant at a more or less uniform rate.

(2) A very slow cell-to-cell movement via the connecting protoplasmic bridges or plasmodesmen combined with a rapid distribution through the plant via the phloem.

There are various methods of examining the dispersal of a plant virus in its host. One is to study the effect of a barrier of dead tissue on the movement of the virus up the stem of some such plant as raspberry or tomato. A ring of tissue including the phloem is killed by the application of steam, chloroform or molten wax. Inoculations below this 'bridge' of dead tissue induce the disease only on that side of the barrier and infection is unable to pass the bridge to the upper portion of the plant.[5, 10] Such experiments show that the virus does not normally move in the dead wood (xylem). If the virus is experimentally injected into the xylem it cannot escape therefrom unless the vessels are ruptured.[10a]

Another method of study is to break the continuity of the phloem by means of a combination of external ringing and removal of the internal phloem with a cork borer. This experiment has been performed with the virus of curly-top of sugar-beet upon the tobacco plant and the results show that dispersal of this virus in tobacco is dependent on the presence of continuous phloem elements.[6]

In a recent study upon the movement of tobacco

mosaic virus within the tomato plant [65] three main methods were used :

(1) Cutting up the stem at various times following inoculation and rooting the portions as cuttings ; the cuttings were then grown until the presence or absence of mosaic could be determined from the symptoms on the new leaves.

(2) Cutting the inoculated plants into sections, grinding these up and using the juice to inoculate seedlings of tobacco or tomato, the presence or absence of virus being judged according to the systemic infections which resulted.

(3) Cutting into sections as for (2) but inoculating on to *N. glutinosa* leaves and judging presence and concentration of virus from the number of local lesions produced.

The first method is the most reliable since the smallest amount of virus would multiply in the cutting, whereas it is by no means certain that a minute quantity of virus might not be missed in the inoculation methods of (2) and (3). Method (1) may be varied by keeping the cut portion of stem in sterile test tubes containing moist cotton-wool for about 10 days to allow of multiplication of the virus instead of planting as cuttings. The pieces may then be ground up and inoculated to the test plants.

Experiments upon the above lines have shown that there is no movement of tobacco mosaic virus from the inoculated leaf for a period of 3–4 days, this period being governed by the growth rate of the plant. The virus then passes out of the inoculated leaf and travels rapidly to the roots of the plant ; about a day later it travels with equal rapidity to the top of the plant. There does not seem to be a continuous flow of virus through the stem in the early stages of infection. This is demonstrated by the occasional absence of virus in successive samples taken from the stem and planted as cuttings ; such virus-free samples occur irregularly between portions containing infection. The presence of developing fruit trusses on the stem seems to influence the movement of the virus, causing it to travel upwards as far as the trusses at the same time that the initial downward movement is occurring and also to enter developing fruits simultaneously with its movement through the stem, although adjacent leaves may remain uninfected for several days or even longer. In pot plants, after the initial rapid infection of the developing leaves at the top, the more mature leaves become successively invaded from the top downwards and from the bottom upwards until the plant is completely invaded by the virus. Complete invasion occurs very quickly in small vigorously growing plants and may take several weeks or longer in larger plants. Large field plants of tobacco or tomato bearing a number of mature leaves never become completely invaded by the tobacco mosaic virus.

The movement of a virus as above described is represented diagrammatically in Fig. 6.

It is considered that these facts favour the theory of a slow cell-to-cell movement of the virus via the plasmodesmen, combined with a rapid distribution through the plant via the phloem.[65]

It has been previously pointed out that some viruses are not sap-inoculable, and in the case of one such virus, curly-top of sugar-beet, there can be no movement by cell-to-cell diffusion because the virus is apparently able to exist only in the phloem to which it is introduced directly by its insect vector. Curly-top can presumably only be produced by artificial inoculation when the virus is injected into the phloem, and this is attended by certain mechanical difficulties.[6]

Studies on the *rate* of virus movement in affected plants indicate a wide range of variation in this direction. The range extends from a rate of 1–2 inches per day in tomato mosaic virus to a rate of 14 inches per

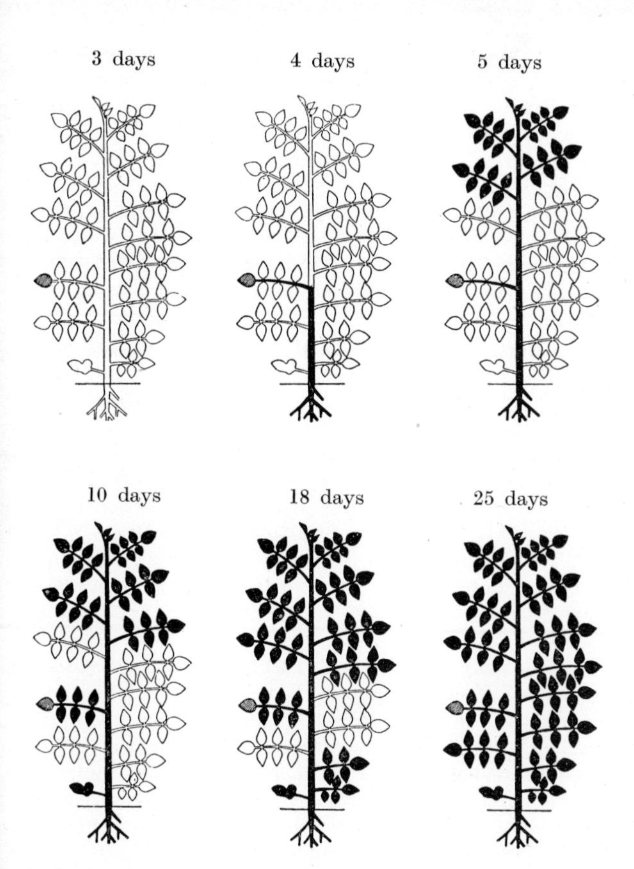

FIG. 6.—Diagram to show the progress of the spread of mosaic (in black) through a medium young tomato plant. Based on tests of Dwarf Champion tomato plants about 15 inches high, growing in 6-inch pots in an unheated greenhouse. Inoculated leaflet shaded. (*After* Samuel)

day for the virus of curly-top. The movement of the curly-top virus in tobacco is relatively slow as compared with the movement in sugar-beet. The fastest movement observed in tobacco was downward from the point of inoculation at the top of the plant to a point 24 inches below in 48 hours ; a rate of $\frac{1}{2}$ inch per hour. In sugar-beet the virus moves much more rapidly. At air temperatures of approximately 85°, 110° and 135° F., the virus moved outward in cotyledons from the point of inoculation a distance of 1 inch in 2 minutes. In larger beets the virus moved downward from the point of inoculation at the distal end of a leaf to a point 6 inches below in 6 minutes, a rate of movement of 60 inches per hour. These rapid movements of virus evidently occur in the phloem and it is suggested that they indicate a rapid translocation of certain plant materials. For these reasons virus may prove useful as an indicator in studies on the movement of elaborated foods.[6]

The movement of tobacco mosaic virus in the host independently of the translocation and transpiration streams has also been studied.[85] The experiments were planned to measure the time necessary for the virus to pass from the upper to the lower epidermis of leaves of *Nicotiana sylvestris*. After a leaf has been inoculated on the upper surface, the virus multiplies in the epidermal cells and may be detected in the mesophyll below in as short a period as 4 hours. In the latter tissue, which comprises about five layers of palisade and spongy parenchyma, the virus continues its course, and at the end of about 30 hours (or 36 hours after inoculation) it has passed into the lower epidermis. Since the thickness of the lamina varies from 250 to 300 μ near the midrib and about midway between the base and the tip of a well-developed leaf of *Nicotiana sylvestris*, this would indicate that the virus travels at an average rate of approximately 7–8 μ per hour through a tissue comprising five layers of cells of the mesophyll and one each of the lower and upper epidermis. It is probable that this slow rate of movement of the virus is by diffusion from cell to cell and may well be the mechanism by which the virus reaches the phloem in which it is mainly carried about the plant.

CHAPTER V

THE VIRUS OUTSIDE THE HOST

Physical Properties

THE reactions of plant viruses in extracted sap to different chemical and physical agents vary over a wide range. Viruses of the tobacco mosaic type will withstand alcohol for some hours up to concentrations of 90 per cent. Potato mosaic viruses are less resistant and are inactivated by 75–80 per cent alcohol, while the virus of beanmosaic is relatively intolerant, being unable to withstand exposure to 25 per cent alcohol for 30 minutes. Perhaps the most resistant virus so far as tolerance to alcohol is concerned is that of tobacco necrosis recently described,[75] which remains viable in 99 per cent alcohol for 48 hours or longer.

Few plant viruses are able to withstand formaldehyde. Some of the potato mosaic group are destroyed by 1 : 500 acting for 2 hours at 27° C.; tobacco necrosis virus, however, seems unusually resistant to the action of formaldehyde.

Viruses as a whole are relatively resistant to the action of glycerine, and this is particularly true of the animal viruses. Such resistance constitutes a point of difference from the behaviour of bacteria which are more intolerant of glycerine.

The reaction of enzymes upon plant viruses is an interesting and important study as it is likely to indicate whether or not these agents are protein in nature. Most of the work which has been done concerns the action of trypsin and pepsin. These two enzymes offer opportunity for comparative study on their reactions with viruses because pepsin is able to digest both native and denatured proteins while trypsin usually fails to digest native proteins. If, therefore, a virus is digested or split by the enzyme, reactivation by heat would be improbable, while if inactivation is due to other causes, heating might induce activity once more. Of the two enzymes, therefore, pepsin is the more likely to produce a permanent inactivation of the virus. Trypsin does actually produce an inactivation effect upon the viruses of tobacco mosaic and tomato yellow (aucuba) mosaic.[11, 46] It is considered, however, for the following reasons that this inactivation is not due to digestion of the virus by the trypsin. Firstly the loss in infectivity of the virus is immediate, secondly it takes place over a wide range of hydrogen-ion concentrations, including some at which trypsin is inactive proteolytically, and thirdly the infectivity of the virus may be regained by heat, by dilution, or by digestion and removal of the trypsin. Some interesting evidence indicates that the trypsin exercises a virus-inhibitory effect upon the plant and that consequently part at least of the apparent inactivation of the virus by trypsin is due to this fact. Trypsin-virus suspensions of tobacco mosaic produce many lesions on *N. glutinosa* but not on the bean (*Phaseolus vulgaris*), although both plants produce equally numerous local lesions on inoculation with normal suspensions of this virus. Furthermore, spraying the test plants with trypsin before inoculation resulted in a marked decrease in the number of local lesions formed subsequently. This inhibitory action is only temporary, since it can be removed by washing the leaves of the plant before inoculation.[78]

Until recently, pepsin has been considered to have no proteolytic action upon the virus of tobacco mosaic and related strains,[11, 46] but this view has recently been challenged.

The infectivity of the virus *is* lost on digestion with pepsin, but only under conditions favourable for proteolytic activity. Pepsin has no appreciable immediate effect on the infectivity of tobacco mosaic virus or its strains, including aucuba mosaic and one or two others, at *p*H 3 to 8 inclusive, as measured by the local lesion method on *N. glutinosa*. Inactivation sets in rapidly, however, on digestion with pepsin at *p*H 3 at a temperature of 37° C., and the rate of inactivation is proportional to the concentration and activity of pepsin and to the time of digestion. Since pepsin inactivates virus only under conditions favourable for proteolytic activity and since the rate of inactivation of virus varies directly with the concentration of active pepsin, it is concluded that the inactivation of virus is due to the proteolytic action of pepsin. This suggests that the virus of tobacco mosaic is a protein, or very closely associated with a protein, which may be hydrolysed with pepsin.[79]

Plant viruses show their greatest variation in behaviour in regard to their reactions to ageing either in extracted sap or in dried tissue. The virus of ordinary tobacco mosaic (tobacco virus 1) remains viable for periods of years both in sterile sap and in the dried tissue of infected plants. On the other hand the virus of tomato spotted wilt loses infective powers in expressed sap in 4 hours at room temperatures; it does not withstand desiccation. In their resistance to ageing the other plant viruses fall within these two extremes, for example potato virus X remains infective in sterile sap for upwards of 3 months at room temperatures while potato virus Y retains viability for only 24–28 hours under similar conditions.

The length of 'life' of many plant viruses in extracted sap may be greatly prolonged by subjecting them to low temperatures; virus 'Y', for example, may be kept viable in sap for so long as 141 days at temperatures of − 10° C.

A probable reason for the rapid inactivation of some plant viruses is the change induced in the extracted sap by oxidation processes. There is a good deal of evidence to support this theory. Extracts of tobacco mosaic in soil when treated with air and oxygen lose much of their activity, while the animal virus of herpes, if kept in open tubes, becomes inactive after 17–24 days, but if sealed with the reducing agent cysteine hydrochloride is still active after this period.[94] In studying the inactivation rate of a virus like that of tomato spotted wilt the course of the reaction can be followed by the primary lesion method and curves can be plotted showing the changes in concentration of the virus with time.

If a suspension of this virus is stirred in the process of inoculation it loses its virulence more rapidly than inoculum standing undisturbed. Again, the rate of inactivation is increased by bubbling air or oxygen through the virus suspension or by the addition of certain oxidizing agents. On the other hand the addition of the reducing agent, sodium sulphite, appreciably retards inactivation.[3]

It has been shown with regard to some animal viruses that they are highly sensitive to the photodynamic action of methylene blue, being inactivated within a few minutes under suitable illumination. The procedure consists in exposing about 10 c.c. of the virus-dye mixture at *p*H 5·8–6 in Petri dishes at a distance of about 2 feet from a 500-watt lamp. This inactivation appears to be an oxidative process depending on the presence of free oxygen.[56] The same experiment has been performed in the study of plant viruses. It was found that the viruses of maize streak and tobacco ringspot were inactivated, tomato streak virus was slightly reduced in virulence, while tobacco mosaic viruses 1 and 6 appeared to be unaffected. While this behaviour is of the same order as that exhibited by the animal viruses, it would appear that the plant viruses are slightly more resistant to the photodynamic action of dyes.[7, 81]

The extent to which a virus may be diluted without entirely losing its virulence depends on its initial concentration. The dilution end-point for a sand-and-pulp filtrate will of course differ from that of unfiltered virus sap, and similarly the end-point of sap from young infected leaves will not be the same as that of sap from old infected leaves, while the species of host plant used also appears to affect the end-point. Allowing for these variable factors, however, there still exists a wide variation in the dilution end-points of the different plant viruses. Tobacco virus 1 seems able to withstand a dilution of one in a million, while certain of the potato mosaic viruses begin to lose infective powers at 1 in 200. The end-point of potato virus X is about 1 in 50,000, while potato virus Y rarely gives infection at greater than 1 in 1,000.

All plant viruses are destroyed at comparatively low temperatures, though here again there exists a variation in the observed thermal death-points. Tobacco virus 1 appears to have the greatest resistance to heat, the death-point being about 90° C., at 10-minute exposures. This is the highest thermal death-point known for plant viruses. Others range from 80° C. for dock mosaic to 43° C. and 42° C. for the viruses of cucumber mosaic and tomato spotted wilt; these are the lowest so far recorded. It is probable, however, that the thermal death-points of plant viruses are not fixed values but depend to a certain extent on the virus content of the suspension.[86]

The response of plant viruses to the action of ultra-violet light is similar to that of animal viruses. Some recent work on the comparative irradiation of tobacco mosaic virus and certain bacteria seems to show that the virus is very resistant to ultra-violet light at 0° C., and is also more resistant than the vegetative stages and spore forms of these bacteria. The curves in Fig. 7 show that while the sensitivity, relative to wave-

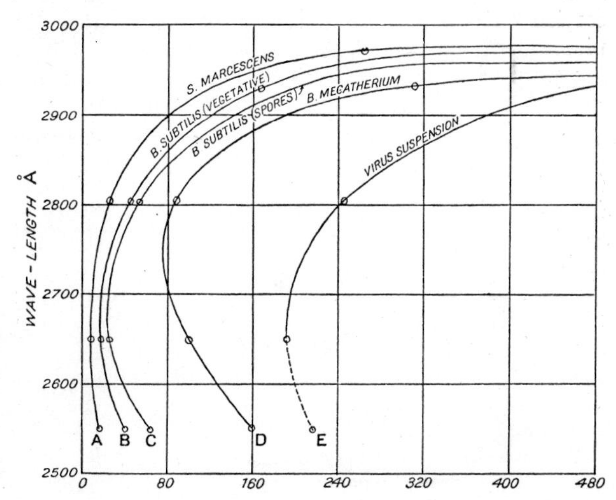

FIG. 7.—Relation between Bacteria (vegetative and spore stages) and Virus at 50 per cent survivor ratio, showing close resemblance in sensitivity relative to wave-length, but entirely different energy magnitudes. (*After* Duggar *and* Hollaender)

length, of the bacteria and the virus is similar, the energy magnitudes are entirely different. The resistance ratio of virus to bacteria is considered to be about 200 : 1.[20]

TISSUE CULTURE

Although the culture of viruses in pieces of living tissue is not, strictly speaking, 'outside the host', it is dealt with for convenience in this section. During the last decade several of the animal viruses have been grown in cultures to which living tissues have been added; examples of these are the viruses of herpes, vaccinia, fowl-pox, the common cold and others. It is only recently, however, that any work has been carried out upon the cultivation of plant viruses in pieces of living tissue. A method has been evolved for producing apparently unlimited growth of the excised root-tips of tomatoes in a liquid nutrient medium [89] and this has led to experiments upon the cultivation of the viruses of tobacco mosaic and yellow (aucuba) mosaic of tomato in isolated root-tips grown *in vitro*. These two viruses will multiply actively in the isolated root-tips for periods of 25–30 weeks, and it is presumed that they can be maintained indefinitely. In obtaining virus-infected root-tips for *in vitro* cultivation it is necessary to grow the roots from systemically infected plants; it does not appear possible to infect the roots by inoculation nor does the virus appear to escape from the roots into the surrounding medium.

The conditions under which the viruses multiply in these cultures can be easily controlled and the method seems favourable for studying the effect of temperature and light on the growth of viruses.[90]

Up to the present time no plant virus has been cultivated in a cell-free medium. The claim was once made that tomato mosaic virus had been induced to multiply in sterile normal tomato juice as far as the twelfth sub-culture. A number of other workers attempted without success to repeat this work and the original claim was later withdrawn. The situation regarding the cultivation of animal viruses in cell-free media is not very clear as there are two opinions on the matter. One school claims that certain viruses such as vaccinia can be so cultivated while the other inclines to the belief that this has not been demonstrated. The great difficulty experienced in attempts to cultivate viruses may be due to the fact that they are obligate parasites incapable of multiplication in the absence of living susceptible cells.

Particle-size of Viruses

It would seem to be a difficult task to apportion a measurable size to virus particles, but this can be done in various ways, as, for instance, by ultra-violet light photography and by ultra-filtration. The only work on the photography of viruses has been carried out by Mr. J. E. Barnard, of the National Institute of Medical Research at Hampstead, and this has been mostly concerned with animal viruses. Some preliminary work, however, has been done by Mr. Barnard with potato virus X and so far as it goes indicates that the particle size lies somewhere in the region of 70 mμ. It is interesting to find that there exists a tendency for the virus particles to combine in aggregates of about 200 mμ in diameter.

The second method of measuring the particle-size of viruses, i.e. by ultra-filtration, is based on the special technique of Elford,[23] who has devised a method of preparing filter membranes which are products of a graded coagulation of ether-alcohol collodion. The important fact about these new membranes is their uniform porosity, which differs markedly from the structure of acetic acid-collodion membranes in which the pore size in any given membrane may extend between fairly wide limits.

The technique of measuring virus particle-size by filtration through these membranes is briefly as follows. After the membrane has been made and graded by adjustment of the ingredients to the approximate pore size required, it must be accurately calibrated. This is done by measuring the rate of flow of a known quantity of water through the membrane and also by determining its water content. The formula for obtaining the average pore size of each grade of membrane is based upon the assumption that

Poiseuille's law governs the rate of flow of water through the pores of the membranes and is as follows :

$$r = 2l\sqrt{\frac{2q\eta}{pv}}$$

where r = radius of pores in cm.
l = length of capillary = thickness of membrane in cm.
q = volume of water passing in c.c./secs.
η = viscosity of water at temperature of measurement in C.G.S. units.
p = pressure producing flow in dynes/sq. cm.
v = total volume of pores (assumed equivalent to water content of membrane).

The pore-size of the membrane which just stops a given virus is known as the filtration end-point and the particle-size of the virus is calculated as a fraction of this pore size according to the following table. This relationship between particle-size and pore-size is based by Elford [24] on experimental evidence in conjunction with the theoretical expectations.

Membrane Average Pore Diameter. mμ	Size of Retained Particle.
10–100	(0·33–0·5)d
100–500	(0·5 –0·75)d
500–1,000	(0·75–1·0)d

d = average pore diameter of limiting membrane for optimum filtration conditions.

By means of this technique it has been found that while the viruses differ widely in their particle-size, the size of each individual virus is extremely uniform. The average particle-sizes of a few representative animal and plant viruses as obtained by this technique are given below. It should be pointed out that there is close agreement in the particle-size values of the same virus as obtained by ultra-violet light photography, and filtration through collodion membranes.

Vaccinia virus 125–175 mμ by filtration, 170–180 mμ by ultra-violet light photography.
Herpes virus 100–150 mμ by filtration.
Foot-and-mouth disease virus 8–12 mμ by filtration.
Tobacco necrosis 20–30 mμ by filtration.
Oxy-Haemoglobin 3–5 mμ by filtration.
(1 mμ = 1 millionth of a millimetre.)

The successful application of Elford's ultra-filtration technique involves the utmost care in attaining standard conditions for the preparation of the membranes and a most painstaking attention to ensure their uniformity.

It is possible that a means of classification of plant viruses by their particle-size can be evolved from the application of the above ultra-filtration technique.

Purification of Virus Suspensions

The question of the purification of plant virus suspensions is a very important one so far as studies upon the physical properties and especially ultra-filtration of a virus are concerned.

There are several methods for freeing the virus suspension of plant proteins and other materials present in the virus sap, but space allows of the mention of only two. One method has been applied particularly to the purification of tobacco virus 1 and consists in precipitating the virus from the juice of virus-diseased plants by means of an aqueous solution of safranin. This precipitate brings down practically all the virus. The virus is apparently held in an inactive condition in the precipitate but is released when the safranin is removed by means of amyl alcohol. It is possible by means of this technique to obtain a water-clear suspension of tobacco mosaic virus with a high virus content.[87]

The second method, which is a modification of one used for purifying a water-soluble ferment, has so far only been used in the purification of potato virus X. It consists, briefly, in precipitating the proteins by means of carbon dioxide which is bubbled through the virus suspension at two different temperatures. The first precipitate formed by the carbon dioxide at 0° C. is discarded and the supernatant fluid is diluted with distilled water at 35° C. Carbon dioxide is passed again through the diluted suspension at this temperature and the precipitate is this time centrifuged down and dissolved in distilled water of an equal volume to the original amount of sap. The supernatant fluid is discarded. The final result is a colourless, slightly opalescent fluid with a very low protein, but high virus, content.[47]

CHAPTER VI

THE VIRUS IN THE INSECT VECTOR

THE relationship between insects and plant viruses forms an interesting chapter in the study of these disease agents and offers scope for fruitful research. It has been pointed out in Chapter III that the majority of insect vectors are to be found in a sub-division (Homoptera) of the plant-sucking insects (Hemiptera), and it may be worth while to speculate on the reasons of this restriction of transmitting power to a particular type of insect. Both by their food and by their method of obtaining it, the Hemiptera are the most likely insects to act as vectors of disease agents. They obtain their food, the sap, by means of a long delicate sucking beak, an ideal injection apparatus, which is thrust into the plant-tissue. This beak contains two parallel channels, down one of which flows the saliva which mixes with the sap in the plant, while up the other flows a mixture of sap and saliva drawn upwards by a muscular pharyngeal pump situated in the head (see Fig. 8). The saliva contains digestive enzymes which dissolve the starch in the plant cells.[92]

In sucking up the sap of a virus-diseased plant the insect naturally draws the virus up also, and this finds its way back to the saliva with which it is discharged into other, and possibly healthy, plants which thereby become infected. Furthermore, most of these insects tap the phloem in search of their sustenance and in so doing inject the virus directly into an area most suit-

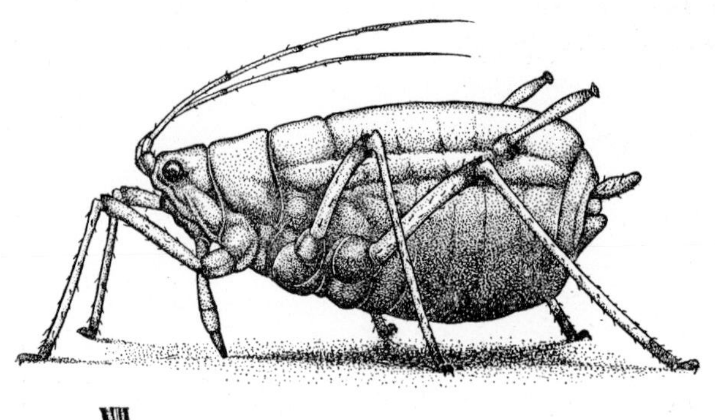

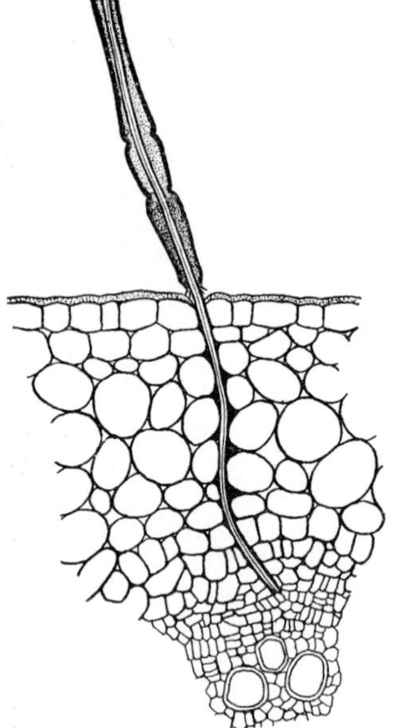

Fig. 8.—Drawing of the aphis, *Myzus persicae* Sulz., in the act of feeding. Note the intercellular path followed by the stylets through the plant tissue to reach the phloem objective. The 'stylet sheath', mainly formed from the insect's secretions, is shown in black surrounding the stylets

able both for its multiplication and its rapid distribution about the plant.

It will be well at this point to consider what is likely to happen to a virus once it has been swallowed by an Hemipterous insect vector, and to understand this a few facts concerning the internal anatomy of such an insect must be given. When the virus is swallowed by a plant-sucking insect it passes into the gut and from there presumably it diffuses through the wall of the alimentary canal into the blood, whence it reaches the salivary glands. Now the alimentary canal of the majority of insects is lined internally by a delicate chitinous membrane which is permeable to digestive enzymes and to the products of digestion. This peritrophic membrane, as it is called, is absent in the Hemiptera, and since it is impermeable to large colloidal particles [92] its absence may well have a bearing on the success of this group of insects as vectors of viruses. If the peritrophic membrane was present in the Hemiptera, the virus on entering might be prevented from passing through the gut walls into the blood and so to the salivary glands but would be conducted straight to the exterior.

It has been shown by experiments with the streak disease of maize that there exist two distinct races of the insect vector of that virus, the leaf-hopper *Cicadulina mbila*. There is no visible difference between these two races which are both apparently the same species. The difference lies in the fact that one race can transmit the streak virus (active) while the other race is unable to do so (inactive). Now Storey [80] has shown that if the wall of the alimentary canal be punctured by a fine needle either just before or just after the insect has fed on a streak-diseased plant, the inactive insect becomes an active one; in other words, it is now able to transmit the virus. This seems to show that for some reason the virus is unable to diffuse through the walls of the gut in the inactive insect and to reach the salivary glands. That the insect is actually imbibing the virus is shown by the recovery of the infective agent from the faeces. On the face of it, therefore, it would appear that permeability or otherwise of the gut wall may play a part in determining the ability of an insect to transmit a plant virus, although there are other factors as well. In many Homopterous insects—the chief vectors of plant viruses—there is a special modification in the digestive system to deal with the excess, due to their mode of feeding, of water and sugars imbibed, and this modification may also have a bearing on virus transmission. Instead of the superfluous fluid being taken into the blood and then eliminated by the Malpighian

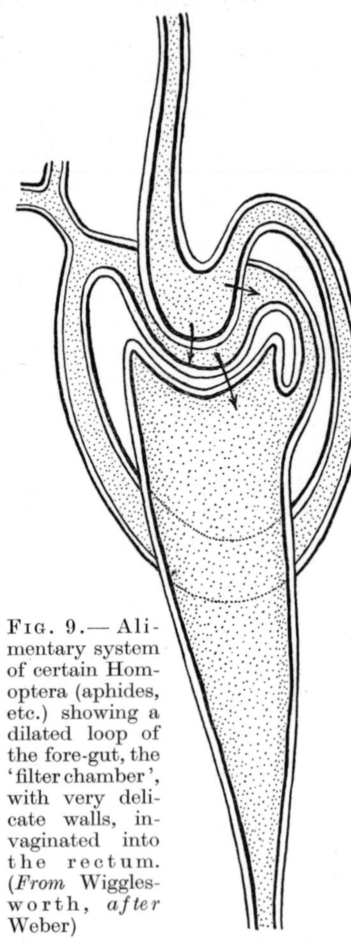

FIG. 9.—Alimentary system of certain Homoptera (aphides, etc.) showing a dilated loop of the fore-gut, the 'filter chamber', with very delicate walls, invaginated into the rectum. (*From* Wigglesworth, *after* Weber)

tubules, a dilated loop of the fore-gut, the 'filter chamber', with very delicate walls, is invaginated into the rectum (Fig. 9). The unwanted fluids are here absorbed or perhaps filtered directly into the hind gut and so discharged.[92] Here again the permeability of this filter chamber may affect the insect's power to transmit a virus. These questions lead naturally to a consideration of certain physical properties of viruses. For example, is there any correlation between the ability of a virus to pass a fine collodion membrane, or in other words its particle-size, and ability to pass through the gut walls or filter chambers of insect vectors? At the present moment there is insufficient evidence to answer this question. It can only be suggested that very small viruses such as those of tobacco mosaic and tobacco necrosis are not easily borne by insects. It may be that they pass rapidly down the alimentary canal and out through the filter chamber without diffusing through the gut wall. Here it should be mentioned that no one seems to have tried to find out whether the virus of tobacco mosaic is present in the 'honey dew' excretion of aphides. The honey dew is formed from the excess liquid passed through the filter chamber and aphides do not easily transmit the virus of tobacco mosaic, at all events from and to the tobacco plant.

It may well be, of course, that ability on the part of an insect to transmit a virus is also partially dependent on some entirely different set of factors, the enzymes present in the salivary fluids, for instance. Although the proteolytic enzymes of insects have not yet been studied on modern lines, it appears that they are more or less like pancreatic trypsin and not of the pepsin type acting in acid media.[92] It has been shown in Chapter V that trypsin apparently does not digest viruses, while pepsin does do so.

There is a very numerous group of plant-sucking insects, the capsid bugs, which as a class have practically no ability to transmit plant viruses although they feed in a precisely similar manner to aphides. This inability to act as vectors is probably connected with the salivary secretions. It is possible that the enzymes in the saliva of capsid bugs are unsuitable for viruses, but there seems to be another reason as well why these insects are not vectors. A virus must have a living cell in which to multiply and this condition cannot be fulfilled for a virus injected by the majority of capsid bugs when feeding. Their saliva is extremely toxic by its own nature to plant cells and each feeding puncture is marked by a patch of dead tissue. The virus, therefore, would be cut off by this necrotic mass from entering the surrounding healthy cells. A very similar phenomenon is often shown during inoculation experiments with necrotic viruses such as that of tomato spotted wilt. Under certain conditions this virus, which is normally systemic in tobacco plants, may find itself isolated by the necrosis of its own primary lesions and is thus prevented from passing over the barrier of dead cells and invading the whole plant.

The exact area of tissue tapped by insects during their feeding may have a bearing on ability to transmit a virus. Some recent work on the curly-top disease of sugar-beet supports this suggestion. It is thought that the virus of this disease is not only confined to the phloem elements but is unable to exist in the cells of the parenchymatous tissue, the contents of which appear to be actually toxic to it. To induce infection, therefore, in a healthy sugar-beet plant the curly-top virus must be introduced directly into the phloem ; this would account for the great difficulty in infecting a plant by mechanical inoculation with the curly-top virus.[6] The insect vector, however, *Eutettix tenellus*, is a phloem-feeder, and thus fulfils the conditions for successful infection. Further, it is suggested that the so-called ' stylet track ' protects the virus from the toxic action of the other tissues *en route* to the phloem. This stylet track, which can be demonstrated by safranin staining in the tissue punctures made by most Hemiptera, appears to be largely of insect origin and contains no plant substance with the possible exception of pectose.

Consideration of the movement of a virus within an insect leads to the question as to whether any increase or multiplication of the agent takes place in the body of the vector. In other words, does an insect once it has imbibed a quantity of virus remain infective without further access to a source of virus, or does it quickly lose power to infect if withheld from further supplies of virus ? A survey of the different vectors shows that both types exist. *Cicadula sexnotata*, the leaf-hopper vector of aster yellows, is capable of retaining the power to infect healthy asters after having lived for two months on a host plant immune to the aster yellows virus. There are several other instances of retention of virus by the insect vector for similar long periods, notably *Eutettix tenellus* and the virus of sugar-beet curly-top, *Aphis rubiphila* and the virus of raspberry leaf curl, *Myzus persicae* and the virus of potato leaf-roll and the thrips *Franklinella insularis* and the virus of tomato spotted wilt. In view of these examples of prolonged retention of virus it is difficult to avoid the conclusion that some multiplication must take place within the insect. Further evidence on this point is afforded by some dilution experiments carried out with the virus of sugar-beet curly-top. It was found that this virus as it occurred in centrifuged infected beetroot juice would stand a dilution of 1 : 1,000, but if extracted from infected beet leaf-hoppers the dilution end-point was 1 : 24,000.[68] Examples of rapid loss of virus by the insect vector are given by the aphis *Myzus persicae* infected with the viruses of cucumber mosaic and potato virus Y respectively. In the first case the aphis loses its ability to infect, immediately after feeding on a healthy plant, and in the second within 24 hours, a period which corresponds to the longevity of the virus in extracted sap. In these two examples there appears to be no multiplication of the virus within the insect.

An important point both from a scientific and economic aspect is the question of inheritance of a virus by the progeny of an infected insect vector. This question has been studied in a large number of cases with negative results. It was affirmed in 1918 that the virus of spinach blight (Cucumber mosaic) was inherited by the progeny of the aphis vector (*Macrosiphum gei*) up to the fourth generation. This has now been disproved.[30a] Recently, however, the claim has again been put forward that a plant virus can be transmitted through the eggs of the infected parent insect.[26] The work has been carried out on the dwarf disease of rice and the leaf-hopper which transmits it (*Nephotettix apicalis* Motsch. var. *cincticeps* Uhl.). Tests were made with the progeny of the following crosses, infective female with infective male, infective female with non-infective male and non-infective female with infective male. From the published data it appears to be essential that the female parent should be infective for the progeny to inherit the virus. It is stated that the majority of the progeny of the other two crosses proved infective although all the individuals of one batch of nymphs from the same parent were not necessarily carrying the virus. This appears to indicate that not all the ova produced in one ovary are always affected.

The above is now the only record of plant virus inheritance in insects by the progeny of an infected parent.

In the case of several plant viruses the specific insect vectors do not become immediately infective to a healthy plant after feeding upon a source of virus infection. Instead there ensues a delay in the development of infective power which may vary from several hours to several days in different insects. Suggestions may be made to account for this though the evidence is not yet sufficient to say which, if any, is the correct one. The virus after being swallowed by the insect would necessarily take some time to pass through the gut wall into the blood and back to the salivary glands before it could be ejected again, and this period might account for the delays up to about 48 hours; it can, however, hardly be the reason for delays of 7–10 days, such as occur with the vectors of tomato spotted wilt and aster yellows. Again, these delays may be due to a necessity for the virus to multiply in the body of the insect, either in the salivary glands or elsewhere, before an infective 'dose' can be produced, in other words, an 'incubation period'. This of course supposes that a virus can multiply inside the vector and there is a good deal of evidence in favour of this theory. The last suggestion raises the possibility of some sort of obligate connexion between virus and insect which would involve a phase of a developmental cycle within the insect, a process comparable to the development of certain protozoan parasites in mosquitoes. This in the writer's opinion is the least likely theory and so far there is very little evidence to support it.

The foregoing facts will have demonstrated that in most cases the relationship between insect and plant virus is by no means a casual or mechanical one. In other words, a virus is not transmitted by any chance insect which may happen to feed on the diseased plant and then migrate to and feed upon a healthy susceptible plant. There is in fact a marked specificity in vector or type of vector with many plant viruses.

This relationship which appears to be obligate in certain cases may depend on a variety of factors.

That there is a very delicate adjustment between the successful vector and the virus is evident when it is considered how one species of insect can transmit a virus while another closely similar species cannot do so. In determining whether an insect can or cannot act as a vector, such factors as method of feeding, *p*H of salivary secretion and the enzymes present therein, permeability of the walls of the alimentary canal and possibly other anatomical characteristics probably all play their part.

CHAPTER VII

IMMUNITY

CERTAIN species of plants may show a natural immunity to a virus which affects other closely related species of plants. *Datura Stramonium* is immune to infection with potato virus Y, although this virus is capable of affecting a large number of other Solanaceae. Furthermore, *D. Stramonium* is very susceptible to infection with another potato virus, i.e. virus X. Reaction to a given virus varies greatly in the different potato varieties, some are very intolerant and become badly diseased while others are tolerant of the virus and carry it without symptoms. A somewhat similar state of affairs can be found with different species of the genus *Physalis* and their reactions with the virus of tobacco mosaic. One species is very intolerant to attack by this virus and develops a severe necrotic disease, a second species shows a mottling disease, a third becomes infected but carries the virus without symptoms, while a fourth appears to be completely resistant.[33]

The only *acquired* immunity in plants to infection by viruses seems to be of the non-sterile type and this has been demonstrated for several virus diseases. Studies with a ringspot virus in America [58] upon different species of *Nicotiana* show that plants of this genus become normally infected with the virus and develop characteristic symptoms. As the plant grows, however, the new leaves show no symptoms and appear healthy except that they are slightly darker in colour. All such 'recovered' plants still retain the virus and sap from them is equally as infective to healthy plants as the sap of plants still showing symptoms. Attempts to re-infect such 'recovered' plants with the ringspot virus are, perhaps naturally, unsuccessful. The foregoing example approximates more to virus 'carrying' than to immunity, particularly as 'recovery' from the same virus under the conditions of lower temperature and light intensity obtaining at Cambridge is very slow and affected plants remain necrotic and obviously diseased for eight or nine months. From this it seems that there may be an element of symptom-masking by temperature concerned in the recovery of plants affected by this ringspot virus.

A more convincing type of immunity is exhibited in certain cases by plants, already infected with an attenuated virus, towards another more virulent strain of the same or a closely related virus. Thus *Datura* plants systemically infected with a mild or G-type of potato virus X are protected against infection by a severe L-type of the same virus.[64] Similarly, cross-immunity studies show that plants infected by attenuated strains of the virus of aucuba or yellow mosaic become immune from unattenuated virus. Plants infected with tobacco-mosaic virus become immune from yellow mosaic, except in the youngest leaves. It has also been shown that plants inoculated with attenuated strains of the virus of tobacco mosaic become immune to infection with both tobacco and yellow mosaic.[43]

There is a good deal of evidence that this immunity is specific, infection by one virus affording no protection against invasion by another virus of a different type. Thus the two viruses described above, those of tobacco and yellow mosaic, offer no immunity to attack by the viruses of tobacco ringspot or cucumber mosaic, and vice versa. Similarly it is possible to inoculate tobacco plants, systemically infected with potato virus X, with the viruses of tobacco necrosis and tomato streak, while on the other hand a plant systemically infected with a green strain of tomato streak is protected against a yellow variant of the same virus.[74] If this specific immunity is proved it should be of value in differentiating related from unrelated viruses (see Chapter VIII).

Little is known regarding the nature of the immunity above described. It may be suggested that certain viruses make use of particular plant proteins and that the virus which enters the plant first is enabled to multiply and to exclude the entrance of another virus which needs similar plant products ; a case, in other words, of first come first served. Alternatively it may be a question of the occupation of particular cells or parts of cells for multiplication. This theory is further borne out by the fact that the second virus may occasionally obtain entrance at one or two points in a ' protected ' leaf, suggesting that a few cells here and there are not invaded by the first virus, thus affording opportunity for the second virus to establish itself at these points. On the other hand it is just possible that protection is due to immunizing substances produced by the infected cells.[43, 82]

Serological Studies

It has been shown by a number of investigators [4, 8, 14, 27, 28, 76] that antibodies reacting specifically with the sap of certain virus-infected plants can be produced by the intraperitoneal injection of rabbits with such expressed saps. The resultant *antibody*, appearing in the blood serum or body fluids of the hyperimmunized animal, reacts specifically with the *antigen* in some observable way. In all experiments up to the present time, the rabbit has been the animal employed for the production of antiserum. Three types of immunologic reactions have been considered, namely, complement-fixation, precipitation and neutralization of the pathogenic properties of the virus.[4] In precipitation the antibody is referred to as *precipitin*. When antigens are mixed with their specific antibodies the mixture has the property of removing the power of normal serum to haemolyse sensitized red corpuscles. This is spoken of as ' complement fixation '.

To obtain the antisera from the rabbits the intraperitoneal injections with the antigen (virus suspension) are made as follows. The quantity of antigen used may be 5 c.cs. and the injections are made at 3- or 4-day intervals. Six to nine days after the last injection the animals are bled, the blood is allowed to clot and the immune serum is collected.

In carrying out experimental work of this nature it must be remembered that the proteins present in normal plant sap are themselves antigenic. Therefore any antigenic property shown by a virus-diseased plant might be due to (1) alteration by the virus of the normal healthy antigenic constituents of the plant, (2) linkage of the virus to the normal healthy constituents of the plant, (3) the virus itself. Now it has been shown by means of precipitin tests that antisera prepared for virus suspensions which have been freed from the normal plant proteins will still react with crude virus-containing plant juice but not with the crude juice of healthy plants. Furthermore, purified preparations of tobacco mosaic virus from tomato plants will react with antisera for crude virus-containing juice from tobacco plants and with the antisera for purified preparations but not with the juice from healthy tobacco plants. These facts show that the reactions secured are due to the virus and not to the plant protein.[8]

Some recent work has shown that plants of tobacco, *Datura* and potato, when infected with potato virus X, contain a common antigen which can be obtained in a relatively pure condition by carbon dioxide precipitation from infected plant saps. This antigen flocculates and fixes complement with the sera of rabbits immunized with either crude sap from infected tobacco plants or with the purified virus suspension (see p. 57), but not with the sera of rabbits immunized with *healthy* tobacco sap or with normal rabbit serum. The anti-virus sera at a dilution of 1 : 10 neutralize the infectivity of purified virus suspensions whilst anti-healthy and normal rabbit serum do not.

The virus antigen is specific to virus X and the closely related potato virus D. It was not found in the sap of tobacco plants infected with the viruses of tobacco mosaic, tobacco ringspot or potato virus Y. No differences were detected between any of the different strains of the X virus used. Sera prepared against one strain reacted equally well with purified suspensions of any other strain. That the antigen is closely associated with the virus was shown by filtration experiments. Filtrates through collodion membranes which were infective flocculated serum, those which were not infective did not do so.[76]

There is further evidence that these precipitin tests are specific and this phenomenon may be used in the classification and differentiation of plant viruses (see p. 81). Thus, extracts of a number of different Solanaceous plants affected with tobacco mosaic, attenuated tobacco mosaic and yellow (aucuba) mosaic all yielded extracts giving a positive precipitin reaction with antiserum to tobacco mosaic (virus 1). On the other hand extracts of plants affected with mosaic diseases other than tobacco mosaic react negatively with antiserum to tobacco virus 1.[4]

The precipitin tests can also be used in the detection of carrier plants, that is plants which, infected with a virus, yet show no symptoms.[4, 28]

Now, as mentioned above, it has been shown that the antibodies produced in rabbits in response to injection with plant virus suspensions have the power of neutralizing or inactivating the suspensions of these viruses. Moreover this neutralizing power is specific in exactly the same manner as the precipitin tests. For example, tobacco-mosaic virus is inactivated only by anti-tobacco mosaic serum, cucumber-mosaic virus only by anti-cucumber mosaic serum, and tobacco ringspot virus only by anti-tobacco ringspot serum. The cross specificity is absolute and the addition to any of these viruses of a heterologous antiserum exhibits no effect. On the other hand the specificity does not extend to distinctions between virus strains even when the strains may induce very different effects in their hosts, as for example the brilliant white, ordinary and symptomless strains of tobacco mosaic. Another interesting point is that it appears possible to differentiate serologically between two viruses which by their symptoms alone may not be distinguished even by the expert. The serological evidence confirms that obtained from a study of host range and other properties in showing that the yellow strain of cucumber mosaic is closely related to the very different-appearing green strain of cucumber mosaic while the same yellow strain of cucumber mosaic is wholly unrelated to a superficially indistinguishable yellow strain of tobacco mosaic.[14]

CHAPTER VIII

THE NATURE OF VIRUSES: CLASSIFICATION

IN trying to arrive at some concept of the nature of plant and animal viruses it must first be realized that they are rather a heterogeneous collection of disease-producing agents and it is not certain that they are all necessarily of the same character. In the animal viruses it is possible to arrange a series with infective particles of diminishing size. Thus at one end of the scale are such agents as psittacosis and vaccinia which appear to consist of bodies like extremely minute organisms and come just within the range of modern microscopical methods. At the other end of the scale are the virus of foot-and-mouth disease and certain of the bacteriophages which approach in size the dimensions of protein molecules in solution. These are not chemical agents in the ordinary sense but it is difficult to picture them as organisms. The agents causing the transmissible tumours of birds fit into this series in many of their properties. Recent work has shown that injection of certain tar products into a chicken will cause the production of a tumour and that a filtered extract from this tumour will reproduce its growth in other chickens from which it can be again passed by filtrates in apparently endless propagation. The filtrates of these tumours have all the properties of living infective agents and not of chemical irritants.[15]

While it cannot be said that plant viruses form an equally heterogeneous group, yet they exhibit a very wide range of behaviour and certain of them appear to be only two or three times larger than the virus of foot-and-mouth disease, the smallest of the animal viruses.

Although a discussion as to whether a virus is a living organism or not is apt to be sterile it may be worth while considering firstly some of those properties of viruses which are usually characteristic of living things, and secondly those points of behaviour which suggest that viruses are of a chemical or inanimate nature. The fundamental attribute of every living thing, that of reproducing itself, is certainly evinced by viruses, but in what manner this multiplication takes place is not known. Apart from multiplication in its host, whether plant or animal, it is possible also to cultivate these agents in living tissue cultures. No one, however, has yet succeeded in cultivating a plant virus in a cell-free medium. Another characteristic of living organisms shown by viruses is *adaptation*. The classic case of adaptation in the animal viruses is the modification in smallpox virus after its passage through various animals. On being passed through monkeys to rabbits and calves and then back to man it is no longer smallpox virus but vaccine virus, and the disease, vaccinia, caused by it is not contagious as is smallpox.[63] While no case is known of a plant virus undergoing quite such a radical change, comparable modifications do occur. The virus of sugar-beet curly-top is reduced in virulence by passage of a certain species of plant so that an attenuated strain is produced. This attenuated or mild type of virus can be reactivated to its original virulence by being passed through a plant of another species. Similarly the virus of tobacco mosaic can be attenuated by heating, but this appears to be an irreversible reaction.

In their reactions to chemical and physical agents viruses do not differ markedly from the visible bacteria. They are all inactivated at fairly low temperatures, they are rapidly destroyed by exposure to ultra-violet light though their resistance in some cases to irradiation seems greater than that of bacteria, and they can be inactivated by most of the usual antiseptics. Although many viruses lose their infective properties on being stored either in dry tissue or sterile plant sap, others may continue viable for long periods, even years, in that state. This property is shared to some extent by certain spore-forming bacteria.

Serological work with viruses, both animal and plant (see Chapter VII), is likely to play an important rôle in discussions on their nature. Specific neutralizing antibodies, complement-fixing antibodies, and antibodies causing flocculation in virus emulsions have been described for both types of viruses.

It is not safe, however, to state that agglutination of particulate matter by a specific serum is proof of autonomous existence in such particles because it has been demonstrated that collodion particles treated with a variety of proteins and then thoroughly washed are specifically agglutinated by the proper antisera.[62]

There are not many properties which suggest that viruses may be of a ‘ non-living ’ chemical nature or products of cellular perversion capable of inciting similar perversions in other cells. There is the extremely small size of some viruses such as those of foot-and-mouth disease and of tobacco mosaic and tobacco necrosis. It is difficult to conceive of these as organisms.

In some ways viruses behave in a manner analogous to that of a chemical substance in that they can be precipitated by safranin, and the claim has been made that tobacco mosaic virus contains no nitrogen. On the other hand recent work with enzymes seems to prove that plant viruses are of a protein nature since they can be digested with pepsin.[79]

On the whole the evidence seems in favour of the theory that some at least of the viruses are of the nature of organisms and here the discussion may well be left until the progress of knowledge allows of more definite conclusions to be drawn.

An interesting point, and one which follows naturally upon the foregoing discussion, is the question how far a virus is a stable entity. This in turn brings up the questions of virus strains and variations in virulence. The present trend of plant virus research is to show that some viruses are not as a rule a single strain but may consist of several closely similar strains. This is well illustrated by potato virus X; when the virus is inoculated to tobacco the symptoms produced consist partly of necrotic rings and partly of a light and dark green mottle. Now, by carefully punching out certain of these different areas in the infected leaf and inoculating them to similar healthy tobacco plants it has been found possible to isolate three or more strains of the virus X. These strains while clearly related to each other yet differ in their virulence to, and the symptoms they produce upon, the tobacco plant.[64] This phenomenon is shown even more clearly by viruses of the tobacco mosaic group. It is quite a common thing for a tobacco plant infected with mosaic to show one or two isolated yellow spots among the light green areas of the mottled leaves. By continuous sub-culturing from these yellow spots it is possible to isolate a virus which produces bright yellow mottling on tobacco instead of the green mottling typical of the ordinary mosaic disease.[37, 49] The same is true of the viruses of cucumber mosaic and tomato streak.[59, 74]

The question which now arises about these virus strains concerns their origin. Were they already present as contaminants in the original virus in high dilution or do they actually arise during the processes of inoculation? If it could be proved that viruses are so unstable as to mutate or produce variant strains during their transfer from plant to plant this would go a long way to explain the immense number of closely similar viruses affecting the potato, the cucumber and tobacco plants. It would also, incidentally, add to the difficulties, already great enough, of evolving a practical scheme of virus classification. What then is the evidence which suggests that a virus mutates during sub-culturing processes? As already described on page 40, inoculation of tobacco mosaic virus to *N. glutinosa* produces local lesions on the inoculated leaf. Now it has been found [37] that when serial transfers were made from these lesions to healthy tobacco plants, the plants developed typical tobacco mosaic. After a time, however, yellow spots again began to develop in these plants. Since some of these yellow strains of mosaic are very much less infectious than the ordinary tobacco mosaic, it is considered unlikely that they can have been carried along through as many as eight or ten serial transfers of the local lesions in *N. glutinosa*. The yellow strains are therefore thought to have arisen anew after the transfer of the purified virus to the tobacco plant. The virus of cucumber mosaic behaves similarly to that of tobacco mosaic and serial transfers have been made with this virus also, using cowpea (*Vigna sinensis*) for the local lesions instead of *N. glutinosa*. The virus of cucumber mosaic produces local symptoms only in cowpea and does not become systemic. However, during serial transfers a strain has been found which produces a systemic infection in cowpea. This strain is thought to have arisen in a primary lesion on that plant.[59] That the strains of a particular virus may differ from each other in more ways than in their symptomatology is shown by some recent work of the writer, who has succeeded in separating certain strains by ultrafiltration.[74]

Some plant viruses can be attenuated or decreased in virulence by treatment with heat and by this means other ' strains ' which appear to be fairly stable can be produced.

It has been shown that an apparently pure stock of tobacco mosaic when incubated in living tissues at temperatures just above 34° C. readily produces attenuated strains.[38]

Recently two closely related but distinct strains have been produced by this method from an apparently pure strain of tobacco mosaic of the ordinary field type. The strains have been described as the mottling and masked strains to distinguish them from the distorting strain from which they have presumably arisen. The masked strain is particularly interesting as it produces no symptoms in tobacco and some other plants, although present in these plants, while the other two strains produce visible symptoms. On the other hand, all three strains produce necrotic lesions in *Nicotiana langsdorffii*. That all three strains are closely related is shown by the many properties which they have in common. The attenuated strains differ, however, from the original stock in being able to increase in host tissues at temperatures high enough to inhibit multiplication of virus of the original distorting strain.[34]

As already mentioned, plant viruses may sometimes be attenuated by passage through some particular host plant and then reactivated to their original virulence by passage of another host plant. Thus, virulent virus of the sugar-beet curly-top disease has been attenuated by passing it through *Chenopodium murale* (the nettle-leaved goosefoot). The attenuated virus remains stable even though passed through successive generations of very susceptible sugar-beets. The attenuated virus can then be reactivated to its original degree of virulence by passage of chickweed

(*Stellaria media*). Virulent virus passed through chickweed remains unchanged.[45]

It is not clear in what way an attenuated strain differs from a virulent strain of the same virus. Presumably it is an intrinsic change in the virus itself, it cannot be merely reduction in the concentration of virus particles, since the final disease resulting from a single particle or group of particles is as severe as one resulting from a heavy dose of virus. The only differences lie in the rapidity of development and possibly severity of the initial symptoms. It may be perhaps that further study of the conditions governing these apparent changes in the virus itself will eventually throw some light upon the nature of these interesting agents.

CLASSIFICATION

Some system of classification of the large group of plant viruses is an urgent need at the present moment and the development of such a system presents very great difficulties. There are two main lines on which a classification could be developed, by cause or effect, in other words classification of the *virus* or the *disease*. From the scientific point of view the first is the most to be desired, and this will probably only be attained after the lapse of some years. A classification of the actual aetiological agents must involve most careful compilation of all the known properties of each virus including its reactions with physical and chemical agents, its insect vectors, serological data, and its symptomatology upon a definite range of host plants. A complicating factor in this classification is the existence of numbers of closely related virus strains, as described in this chapter, and the apparent ability of viruses to produce more strains. For the time being it may be necessary to confine the classification to virus *groups*, thus the potato mosaic viruses can be divided into two classes, the type viruses being known as the X and Y groups. These two types differ widely from each other in their physical properties, insect vectors and symptomatology. The cucumber mosaic viruses also form another distinct group of similar agents.

A start has been made in classifying the individual viruses of the tobacco mosaic group and they have been numbered as separate entities such as tobacco virus 1, 2, 3, &c. How far it will be possible to apply this method of classification to the other groups of viruses remains to be seen.

There are of course certain viruses which by reason of their own characteristics fall quite easily into their place in any system of classification. Such are the viruses of potato leaf-roll, bunchy-top of bananas and spike of sandal. So far as is known at present these diseases are due to single viruses, but the fact that they are not mechanically transmissible has militated against much research into their nature and properties and it is not known whether these viruses exist in numerous similar strains.

From the practical point of view there is much to be said for a classification of virus *diseases*, since it is the virus causing the worst symptoms which will most interest the grower. Virus diseases can be roughly classified into *mosaics*, where the symptoms are mainly of the chlorotic mottling types, *yellows*, where there is no mottling but a uniform chlorosis, *necrotic* diseases, where there is a definitely destructive action on the cells of the plant, and finally a miscellaneous group of disorders where the symptoms are mainly malformation, overgrowth or some other abnormality of growth. From the scientific standpoint a classification of virus diseases breaks down when it is realized firstly that the same virus can cause entirely different diseases on different plant species or even on different varieties of the same species, and secondly that two different viruses may cause diseases in the same plant species so closely similar as to deceive even the expert.

CHAPTER IX

CONTROL

THE question of the control of plant virus diseases can be approached from a number of aspects but the various control methods are not applicable to all virus diseases. It will be more convenient therefore to deal with them seriatim, showing their application in particular cases.

The ruthless eradication of diseased plants from a crop, or 'roguing' as it is usually called, has been practised with some success in the control of potato leaf-roll, peach yellows, bunchy-top of bananas and to a lesser degree in mosaic of raspberries. The method must be employed before the diseases have spread too widely in the crop and it is advisable to remove also those plants immediately adjoining the diseased plant. All plants thus rogued out must be completely destroyed, preferably by burning. In the potato crop roguing is more likely to be successful with a disease such as leaf-roll than with the mosaic group, since the symptoms of leaf-roll are conspicuous and easily recognizable and the virus does not lie latent without symptoms as the mosaic viruses do in some 'carrier' potato varieties. Roguing of commercial raspberry plantations has been found impracticable, but it can be carried out in 'cane nurseries'. Any canes showing symptoms of mosaic should be dug up and burned, together with the canes adjoining on either side. It is important that all roots should be removed and burned and that no rogued plants are left lying in the nursery rows.[29]

It is perhaps hardly necessary to emphasize the importance of using virus-free stocks for the propagation of a crop. Seed should never be saved from any plant which is not free of virus contamination and similarly all cuttings, suckers, tubers or any part of a plant used for vegetative propagation must be above suspicion if a virus-free crop is desired. Since potato-growing affords an excellent example of vegetative propagation on a large scale, the method adopted to ensure the selection of virus-free tubers will be described. '*Indexing*', as it is called, is largely practised in some of the potato-growing areas of the U.S.A.[42] and has also been attempted in this country. The health of the selected tuber stocks is tested by growing a portion of each tuber during the winter under forcing conditions in the glasshouse. The tubers are first numbered with Indian ink or some similar indelible substance and a piece of each tuber correspondingly labelled is then grown and examined for symptoms of virus disease. A convenient method of obtaining a representative portion of a tuber is to extract, by means of a sterile cork borer, a plug of tissue bearing an eye. This method of tuber indexing is used to eliminate the virus-infected tubers and to obtain healthy foundation stocks. The stocks are next multiplied in isolation, at first in the insect-proof glasshouse and later at high altitudes, on the sea coast, or in any locality where the risk of insect-borne contamination is negligible. Small lots of tubers are then parcelled out among growers who can increase them in isolation on their own farms and gradually replace their old stocks. A co-operative effort on these lines has been inaugurated in Wales where a number of progressive growers produce each year in isolation sufficient virus-free seed tubers to serve their own needs.

A word of warning is necessary as regards the

indexing method as above described: it fails to reveal the identity of 'carrier' potato varieties, i.e. those which bear a latent virus within them and are in consequence a source of infection for other healthy non-carrying potato varieties. This difficulty can be avoided to a certain extent by making additional tests of the index shoots by inoculations from them to tobacco and *Datura*. The production of symptoms in these plants indicates the presence of latent virus in the apparently healthy potato shoots.

An important factor in control is the removal of susceptible weeds or other plants which may act not only as reservoirs of virus infection but also, in the case of an annual crop, afford opportunity to the virus for overwintering. The removal of three species of weeds from the neighbourhood of tobacco crops in Java has materially helped to control the leaf-curl disease of that crop. These weeds serve to carry the virus over the period when the tobacco is not in the field.

The importance of removing old diseased plants after the crop has been harvested is well illustrated by the case of cotton leaf-curl in the Sudan. It is customary to separate the successive cotton crops in the Gezira by a 'dead season' lasting from two to three months. Cotton fields are dried out from April onwards, and when the last of the cotton has been harvested the stems are chopped off just above ground level and the roots left in the soil. The whiteflies which are the vectors of the leaf-curl virus also disappear during the dead season. A percentage of the roots and stems of the cotton plants which are left in the soil sprout again, and it has been discovered that this 'ratoon' cotton, as it is called, enables the virus to carry over the period when neither the cotton crop nor the insect are available. Careful attention to the eradication of all ratoon cotton has materially helped to control this serious disease.[1]

Some viruses have an almost unlimited host range and this fact is of great importance and must be considered when devising control measures. The virus of tomato spotted wilt can infect a great number of hothouse plants besides the tomato, notably dahlias, arum lilies and chrysanthemums. Tomatoes should therefore be grown by themselves if possible, and houses where these plants have been grown should be carefully disinfected to kill the thrips vector before the next crop is put in.

Whenever practicable two crops, both susceptible to a prevalent virus, should not be grown in close proximity. In the south-west of France where tobacco is cultivated it is the usual practice to grow a few rows of potatoes alongside the tobacco crop. The result of this practice is infection of a high percentage of the tobacco with the aphis-borne virus Y to which both plants are equally susceptible.

Since the majority of plant viruses are dependent upon insects for their spread in the field, measures against these insect vectors constitute an important part of the control of plant virus diseases. There are several ways of approaching this problem, by direct attack on the insect itself with the aid of chemicals, by the use of parasitic or predatory insects which prey on the insect vectors—this method is more likely to be of practical use in an enclosed space such as a glasshouse—by cultural methods, and by protecting the crop in the seed bed with screens or cages. Two examples of the cultural method of approach to this problem may be quoted. The first concerns the rosette disease of groundnuts in the Gambia, where it has been found that the incidence of the disease can be much reduced by covering the ground with vegetation before the seasonal invasion of the aphis vectors has developed. In the second case the method of ‘ trap cropping ’ has been practised in relation to the leaf-crinkle disease of sugar-beet in Germany. A few marginal rows of sugar-beet are sown two or three weeks in advance of the main crop. These rows are above ground about the time of the appearance of the insect vector, a Tingid bug, which accordingly assembles on them in large numbers. The insects are then destroyed and the infested beets ploughed under.[70]

Some success has been attained in ' warding off ' the insect vector from the seed bed by means of cloth screens or cages. This idea was first developed in the case of ' aster yellows ', the virus of which is transmitted by a leaf-hopper.

Two types of shield have been tested. In one type cloth-covered side walls or ' fences ' 6 feet high with tops uncovered were made. These combined with roguing reduced the incidence of yellows but were not found to be commercially satisfactory under local conditions. In the second type cloth-covered cages or houses were employed. The tops and sides of the enclosures were completely covered with cloth not coarser than 22 by 22 threads per inch. These were found satisfactory for the practical control of the disease.[39]

A similar form of protection has been employed in production of virus-free potato ' seed ' in the U.S.A. Certified tubers were grown under large cages made of the ' aster cloth ' above mentioned. The experiments were made with two cages, covering 32 and 4 square rods respectively. The yield rate under the cloth was 140 barrels (385 bushels) per acre, this being about the same as in the open field. The cloth had about 21 meshes to the inch and cast some shade but not enough to change the appearance of the plants as do small cages of heavier cloth. On the whole the cages seem to have been fairly efficient in excluding the aphis vectors of potato viruses, but the method is likely to prove too expensive for general use, at all events in this country.[25]

The time may not be too far distant when accurate forecasts of insect attacks can be made and this will have its application in the control of virus diseases. A step in this direction has been made in the formulation of an index figure of aphis infestation of potato crops. It has been found that the index figure of infestation at centres in which there was a rapid increase in virus diseases among the potato stocks always exceeded 100 of the chief aphis vector (*Myzus persicae*) per 100 leaves and attained such numbers as 1,000 individuals per 100 leaves. In Fig. 10 are shown the infestations of *Myzus persicae* in successful and unsuccessful seed potato-producing centres.[50]

Resistant Varieties

Perhaps the most promising method of control lies in the production of virus-resistant varieties of plants by breeding on genetic lines and some success in that direction has already been achieved. Several good varieties of mosaic-resistant sugar-cane have been produced, known as the P.O.J. strains, and the substitution of these for susceptible varieties in most of the sugar-growing areas has reduced the disease to one of small importance although at one time it threatened the very existence of the sugar-cane industry. During the last few years the general use in Java of the resistant P.O.J. 2878 has led to the question of sugar-cane mosaic being pushed into the background, but the extreme susceptibility of certain new varieties has once more brought it into prominence.

Much work has been carried out in America in trying to evolve a strain of sugar-beet resistant to the virus of curly-top. By the combination of a number of strains selected for resistance, a variety has been produced which has a fair degree of resistance to this disease and is reasonably satisfactory in other respects. This variety, designated U.S. No. 1, has been submitted to testing trials in various parts of the U.S.A.

during 1930 and 1931. These extensive tests demonstrated that it is markedly superior in resistance to

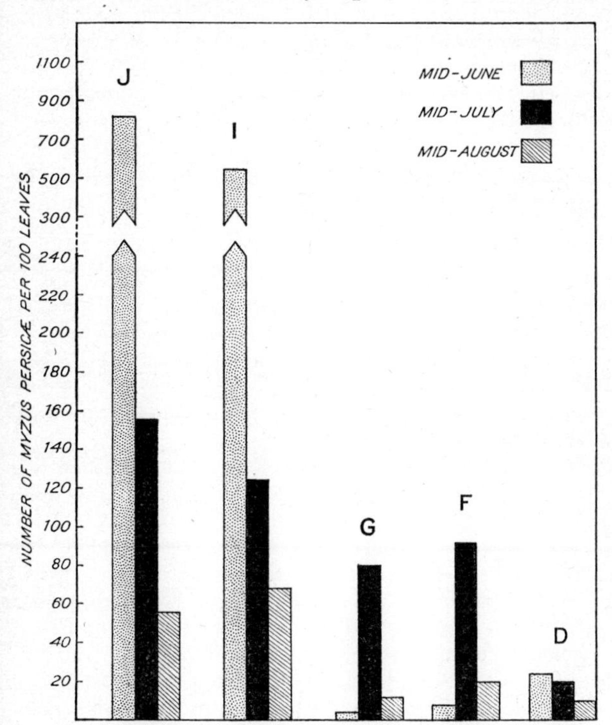

FIG. 10.—Infestation of *Myzus persicae* in 1933 in N. Wales. **J, I**: Unsuccessful seed potato-producing centres. **G, F, D**: Successful seed potato-producing centres. (*After* Maldwyn-Davies)

any of the standard commercial brands with which it was compared.[12]

Two strains of cotton resistant to leaf-curl, designated X1530 and X1730, have been evolved. These are of Sakel type, of outstanding vigour and fruitfulness, and they combine with these qualities a very high degree of resistance to leaf-curl. The following percentage infection figures relating to a chequer-board variety test indicate the degree of resistance to leaf-curl shown by these strains. The figures were obtained 3 months after the disease first appeared.[1]

PERCENTAGE OF PLANTS INFECTED

Sakel Cotton	91·5 per cent
X1530 ,,	3 ,, ,,
X1730 ,,	2 ,, ,,
X1030 ,,	2 ,, ,,

Work on similar lines is proceeding with other crops, notably groundnuts and potatoes,[67] while a variety of tobacco called Ambalema seems resistant to the virus of tobacco mosaic.[54] There seems no reason why virus-resistant varieties of other crops also should not be obtained.

Since some viruses appear able to exist in numerous strains, the possibility must be borne in mind when considering the question of virus-resistant crops that a strain of virus may appear which is capable of attacking a plant normally resistant to that virus.

Where an outbreak of virus disease threatens to destroy an industry by an epidemic of unusual virulence it may be necessary to invoke the aid of legislation and make that particular disorder a notifiable disease. Such a procedure might become advisable with the tomato virus disease known as spotted wilt, which at times causes great loss to the tomato-growing industry.

It has been shown in Chapter VII that inoculation of a plant with a weak strain of virus will protect that plant from invasion by a more virulent strain of the same virus. It may be possible in the future to make practical use of this phenomenon and by means of such protective inoculations render a crop immune from attack. A preliminary attempt in this direction has already been made by inoculating tomato plants with a masked strain of tobacco mosaic (see p. 80) and comparing them respectively with similar plants inoculated with a strain of distorting virus and plants left exposed to natural infection. It was found that the plants inoculated with the masked virus strain were distinctly better in average ripe-fruit weight and slightly better in average yield of green fruit than the set inoculated with distorting-strain virus or the set not inoculated at all, but exposed to chance infections in the garden during the growing season.[34] A word of warning, however, is necessary concerning this protective 'vaccination' of plants. If by chance such plants become infected with another virus, the combination may 'flare up' into a much worse disease than would be caused by either virus acting separately. This is well shown by the tomato virus disease known in America as 'streak', or more correctly 'double virus streak', which is due to the combined effects of tobacco virus 1 and potato virus X. A similar phenomenon occurs with virus diseases of animals, and the occasional production of a serious disease in man after vaccination may possibly be due to the combination with the vaccinia of a latent virus already present.

In conclusion a word may be said on the control of certain virus diseases of the peach. The claim has been made that if young peach trees affected with yellows and rosette are subjected to high temperatures they may be cured of these virus diseases. Experimental cures were obtained by incubating potted trees in a hot room at a temperature varying from 34·0° C. to 36·3° C. for two weeks or longer. Some difficulty was experienced in destroying the virus in the roots which, being in moist earth, do not reach the temperatures to which the tops are exposed owing to evaporation.[44]

CHAPTER X

SOME COMPARISONS BETWEEN THE PLANT AND THE ANIMAL VIRUSES

IT will be of interest to compare the plant viruses with some of the viruses affecting animals. The close similarity of many of their properties will leave no doubt in the reader's mind that these two groups of disease agents are of the same general nature.

Firstly, as regards their particle-size, it has been made possible by the use of Elford's ultrafiltration technique (see p. 55) to gain an accurate estimate of the size of a number of different animal viruses. As this technique is only now being applied to the measurement of plant virus particles, a comprehensive comparison cannot yet be made. There appears to be, however, a similar variation in the particle-sizes of the different plant viruses to that found in the animal viruses though it is not yet known whether the range of variation in the former is equally wide. The smallest animal virus is that of foot-and-mouth disease, the average particle-diameter of which is 8–12 mμ (1 millimetre = 1,000 μ = 1,000,000 mμ). The particle-sizes of some other animal viruses are as follows: vesicular stomatitis virus 70–100 mμ, vaccinia virus 125–175 mμ, herpes virus 100–150 mμ, louping-ill 15–20 mμ, fowl-plague 60–90 mμ, while the bacteriophages range in size from 8–12 mμ to 50–75 mμ. The smallest plant viruses yet measured are those of tobacco mosaic and tobacco necrosis, which have a particle-size of 17–25 mμ and 20–30 mμ respectively. Potato virus X is a larger virus, measuring approximately 80–120 mμ.

In their reactions to physical and chemical agents the two types of viruses are quite comparable. The thermal death-points of plant viruses range from 90° C. to 42° C., while those of the animal viruses range from 70° C. to 45° C.

The resistance of both types of viruses to various chemicals seems to be of the same order although some of the plant viruses may have a greater tolerance to alcohol; the virus of tobacco necrosis, for instance, will remain viable in 99 per cent alcohol for 71 hours.

As regards dilution the animal viruses appear able to withstand slightly higher dilutions than the plant viruses. With the exception of the tobacco mosaic group, which are still infectious at 1 in 1,000,000, the dilution end-point of a large number of plant viruses is below 1 : 10,000. On the other hand the virus of foot-and-mouth disease will stand dilutions of 1 in 10,000,000, and the virus of polyhedral disease of caterpillars is still infectious at 1 in 1,000,000.

Plant viruses vary greatly in their resistance to ageing in extracted sap or dried plant tissue and the length of time they will remain viable under these conditions ranges from a period of years with tobacco mosaic virus in dried tissue to three or four hours with the virus of tomato spotted wilt in extracted sap. Similarly the viruses of fowl-pox and grasserie of silkworms can remain infectious after two years in a dried state, while on the other hand the virus of yellow fever is unable to withstand drying.

The rapid inactivation of some viruses in extracted fluids is thought to be due to oxidative changes and the rate of inactivation for both types of virus can be greatly lessened by the addition of reducing agents and to a less extent by freezing.

It has been shown that many animal viruses are highly sensitive to the photodynamic action of methylene blue,[56] a concentration of 1 part in 100,000 inactivating the viruses within a few minutes under suitable illumination. The same process has been applied in the case of a few plant viruses. It was found that maize streak virus was similarly inactivated,[81] but not the tobacco mosaic viruses 1 and 6[7] or potato virus X.

Exposures to ultra-violet light rays destroy viruses but susceptibility to inactivation in the case of plant viruses seems to be less than that of bacteria (p. 53).

Both types of viruses have affinities with insects though this relationship is closer in the case of the plant viruses. Nevertheless the transmission by insects of such viruses as those causing yellow fever and sand-fly fever seems to be of the same order as the insect transmission of many plant viruses. There is not, however, in the animal viruses such a high degree of specificity of insect vector.

There is one instance where a plant virus is said to be inherited by the progeny of an infective parent, this is dwarf disease of rice and the infective agent is stated to pass to the next generation of the leaf-hopper vector provided that the female parent is infected.[26] Similarly in a disease of the typhus group, known as Tsutsugamushi, in which the vector is a mite *Trombicula akamushi*, the virus appears to be inherited by the progeny of the infected mite.

As already pointed out, some plant viruses consist of a number of closely similar strains, and this state of affairs is also found with the animal viruses. Furthermore in both groups it is possible to induce experimental modifications in existing virus strains.

Intracellular inclusions occur in the tissues of both plants and animals affected with viruses, but it is not at all certain that these inclusions are analogous. The inclusions found in virus-diseased plants (Fig. 5) are thought to be merely aggregations of the cell cytoplasm and so to be the effect rather than the cause of the disease. It cannot be said, however, that the virus is not also present in these cell inclusions. The intracellular inclusions characteristic of animal virus diseases seem to be much more definitely associated with the causal agent, and in certain cases appear to be colonies of the actual virus bodies.

The same difficulty is experienced in cultivating both virus types in a cell-free medium, and this has not yet been achieved with the plant viruses, while opinion is divided on the question as regards the animal viruses. There is no doubt, however, that such cultivation is exceedingly difficult, if not impossible, and it emphasizes the very close relationship which seems to exist between the virus and the living cell.

REFERENCES

General Works on Viruses (Plant and Animal)

ATANASOFF, D. 1934. *Virus Diseases of Plants : A Bibliography*. Houdojnik Printing Co., Sofia.

BEAUVERIE, M–A. 1932. 'Les maladies a ultravirus des plantes.' *Annales du service botanique et agronomique de Tunisie*, Tome IX (published separately).

FAIRBROTHER, R. W. 1934. *Handbook of Filterable Viruses*. William Heinemann (Medical Books), Ltd., London.

FINE, JOSEPH. 1932. *Filterable Virus Diseases in Man*. E. and S. Livingstone, Edinburgh.

GARDNER, A. D. 1931. *Microbes and Ultramicrobes*. Methuen's Monographs on Biological Subjects.

GRAINGER, JOHN. 1934. *Virus Diseases of Plants*. Oxford Univ. Press.

MCKINLEY, E. B. 1929. 'Filterable Virus and Rickettsia Diseases.' *Philippine Journ. Sci.*, **39** (published separately).

OTERO, J. I., and COOK, M. T. 1934. 'Partial Bibliography of Virus Diseases of Plants.' *Journ. Agric. Univ. Puerto Rico*, **18**, 1–2.

RIVERS, THOMAS M. 1928. *Filterable Viruses*. Baillière, Tindall & Cox, London.

SMITH, KENNETH, M. 1933. *Recent Advances in the Study of Plant Viruses*. J. and A. Churchill, London.

SYSTEM OF BACTERIOLOGY. 1930. 'Virus Diseases: Bacteriophage,' **7**. Med. Res. Counc. H.M. Stationery Office, London.

* * * * * *

[1] BAILEY, M. A. 1934. 'Leaf Curl Disease of Cotton in the Sudan.' *Empire Cotton Growing Review*, **11**, 4, 280.

[2] BALD, J. G. 1935. *Investigations on Spotted Wilt of Tomatoes. III. Infection in Field Plots.* (In the press.)

[3] —— and SAMUEL, G. 1934. 'Some Factors affecting the Inactivation Rate of the Virus of Tomato Spotted Wilt.' *Ann. Appl. Biol.*, **21**, 2, 179–90.

[4] Beale, Helen Purdy. 1934. 'The Serum Reactions as an Aid in the Study of Filterable Viruses of Plants.' *Contr. Boyce Thomp. Inst.*, **6**, 3, 407–35.

[5] Bennett, C. W. 1927. 'Virus Diseases of Raspberries.' *Agric. Exp. Sta. Mich. State Coll. Tech. Bull.*, 80.

[6] —— 1934. 'Plant Tissue Relations of the Sugar Beet Curly-top Virus.' *Journ. Agric. Res.*, **48**, 8, 665–701.

[7] Birkeland, J. M. 1934. 'Photodynamic Action of Methylene Blue on Plant Viruses.' *Science*, **80**, 2077, Oct. 19, 357–8.

[8] —— 1934. 'Serological Studies of Plant Viruses.' *Bot. Gaz.*, **95**, 3, 419–36.

[9] Blakeslee, A. F. 1921. *Journ. Genetics*, **11**, 17–36.

[10] Caldwell, J. 1930. 'Movement of Mosaic in the Tomato Plant.' *Ann. Appl. Biol.*, **17**, 3, 429–43.

[10a] —— 1931. 'Further Studies on the Movement of Mosaic in the Tomato Plant.' *Ann. Appl. Biol.*, **18**, 3, 279–98.

[11] —— 1933. 'The Nature of the Virus Agent of Aucuba or Yellow Mosaic of Tomato.' *Ann. Appl. Biol.*, **20**, 100–16.

[12] Carsner, E. 1933. 'Curly-top Resistance in Sugar Beets and Tests of the Resistant Variety U.S. No. 1.' *U.S. Dept. Agric. Tech. Bull.*, 360.

[13] Carter, W. 1930. 'Ecological Studies of the Beet Leafhopper.' *U.S. Dept. Agric. Tech. Bull.*, 206.

[14] Chester, K. S. 1934. 'Specific Quantitative Neutralization of the Viruses of Tobacco Mosaic, Tobacco Ring-spot and Cucumber Mosaic by Immune Sera.' *Phytopath.*, **24**, 11, 1180.

[15] Dale, H. H. 1934. 'Chemical Ideas in Medicine and Biology.' *Science*, **80**, 2077, 343–9.

[16] *Degeneration of the Strawberry*. 1934. Imper. Bur. Fruit Product. Tech. Comm., Feb. 5.

[17] Dobroscky, I. D. 1931. 'Studies on Cranberry False-blossom Disease and its Insect Vector.' *Contrib. Boyce Thomp. Instit.*, **3**, 1, 59–83.

[18] Doolittle, S. P., and Gilbert, W. W. 1919. 'Seed-transmission of Cucurbit Mosaic by the Wild Cucumber.' *Phytopath.*, **9**, 326–7.

[19] Duggar, B. M. 1930. 'Problem of Seed-transmission of Typical Mosaic of Tobacco.' *Journ. Bact.*, **19**, 1, 20.

[20] —— and Hollaender, A. 1934. 'Irradiation of Plant Viruses and of Micro-organisms with Monochromatic Light. II. *Journ. Bact.*, **27**, 3, 241–56.

[21] —— and Johnson, B. 1933. 'Stomatal Infection with the Virus of Typical Tobacco Mosaic.' *Phytopath.*, **23**, 12, 934–48.

[22] L'ECLUSE, CHARLES DE. 1576. *Rariorum Aliquot Stirpium per Hispanias Observatarum Historia.* 529 pp., illus. Antverpiae.

[23] ELFORD, W. J. 1931. 'A New Series of Graded Collodion Membranes.' *Journ. Path and Bact.*, **34**, 505–21.

[24] —— 1933. 'The Principles of Ultrafiltration as Applied in Biological Studies.' *Proc. Roy. Soc. B.*, **112**, 384–406.

[25] FOLSOM, D. 1934. 'Growing Seed Potatoes under an Aster Cloth Cage.' *Amer. Potato Journ.*, March.

[26] FUKUSHI, TEIKICHI. 1933. 'Transmission of Virus through the Eggs of an Insect Vector.' *Proc. Imper. Acad.*, **9**, 8, Bot. Instit., Hokkaido Imper. Univ., Sapporo, Japan.

[27] GRATIA, ANDRÉ. 1933. 'Pluralité, Hétérogénéité, Autonomie Antigéniques des Virus des Plantes et des Bactériophages.' *Comptes Rendus des Séances de la Soc. Belge de Biol.*, **114**, 1382.

[28] —— et MANIL, P. 1934. 'Différenciation Sérologique des Virus X et Y de la Pomme de Terre chez les Plantes infectées ou Porteuses de ces Virus.' *Comptes Rendus des Séances de la Soc. Belge de Biol.*, **117**, 490.

[29] HARRIS, R. V., and GRUBB, N. H. 1933. 'The Commercial Control of Raspberry Mosaic Disease.' *Ann. Rept. East Malling Res. Sta.*, 1932, 149–51.

[30] HENDERSON, R. G. 1931. 'Transmission of Tobacco Ringspot by Seed of Petunia.' *Phytopath.*, **21**, 2, 225–9.

[30a] HOGGAN, I. A. 1933. 'Some Viruses affecting Spinach and Certain Aspects of Insect Transmission.' *Phytopath.*, **23**, 5, 446–74.

[31] HOLMES, F. O. 1928. 'Accuracy in Quantitative Work with Tobacco Mosaic Virus.' *Bot. Gaz.*, **86**, 1, 66–81.

[32] —— 1929. 'Inoculating Methods in Tobacco Mosaic Studies.' *Bot. Gaz.*, **87**, 1, 56–63.

[33] —— 1932. 'Symptoms of Tobacco Mosaic Disease.' *Contrib. Boyce Instit.*, **4**, 3, 323–57.

[34] —— 1934. 'A Masked Strain of Tobacco Mosaic Virus.' *Phytopath.*, **24**, 8, 845.

[35] —— 1934. 'Increase of Tobacco Mosaic Virus in the Absence of Chlorophyll and Light.' *Phytopath.*, **24**, 10, 1125.

[36] HUTCHINS, LEE M. 1933. 'Identification and Control of the Phony Disease of the Peach.' *Office of the State Entomologist, Atlanta, Ga. Bull.*, 78.

[37] JENSEN, J. H. 1933. 'Isolation of Yellow-mosaic Viruses from Plants infected with Tobacco Mosaic.' *Phytopath.*, **28**, 12, 964–74.

[38] JOHNSON, J. 1926. 'The Attenuation of Plant Viruses and the Inactivating Influence of Oxygen.' *Science*, **64**, 210.

[39] JONES, L. R., and RIKER, R. S. 1931. 'Wisconsin Studies on Aster Diseases and their Control.' *Wisconsin Agric. Exp. Station. Res. Bull.*, 111.

[40] KENDRICK, J. B. 1934. 'Cucurbit Mosaic transmitted by Muskmelon Seed.' *Phytopath.*, **24**, 7, 820.

[41] KEUR, J. Y. 1933. 'Seed-transmission of the Virus causing Variegation of Abutilon.' *Phytopath.*, **23**, 1, 20.

[42] KOTILA, J. E. 1931. 'Experiments with the Tuber-index Method of Controlling Virus Diseases of Potatoes.' *Michigan State Coll. Tech. Bull.*, 117.

[43] KUNKEL, L. O. 1934. 'Studies on Acquired Immunity with Tobacco and Aucuba Mosaics.' *Phytopath.*, **24**, 5, 437–66.

[44] —— 1935. 'Heat Treatment for the Cure of Yellows and Rosette of Peach.' Abstr. in *Phytopath.*, **25**, 1, 24.

[45] LACKEY, C. F. 1932. 'Restoration of Virulence of Attenuated Curly-top Virus by Passage through *Stellaria media*.' *Journ. Agric. Res.*, **44**, 10, 755–65.

[46] LOJKIN, M., and VINSON, C. G. 1931. 'Effect of Enzymes upon the Infectivity of the Virus of Tobacco Mosaic.' *Contrib. Boyce Thomp. Instit.*, **3**, 2, 147–62.

[47] MACCLEMENT, D. 1934. 'Purification of Plant Viruses.' *Nature*, **133**, 3368, 760.

[48] MCKAY, M. B., and WARNER, M. F. 1933. 'Historical Sketch of Tulip Mosaic: the Oldest Known Plant Virus Disease.' *Nat. Hortic. Mag.*, **12**, 3, 179–216.

[49] MCKINNEY, H. H. 1926. 'Virus Mixtures that may not be Detected in Young Tobacco Plants.' *Phytopath.*, **16**, 893.

[50] MALDWYN-DAVIES, W. 1934. 'Studies on Aphides infesting the Potato Crop. II. Aphis Survey: its Bearing upon the Selection of Districts for Seed Potato Production.' *Ann. Appl. Biol.*, **21**, 2, 283–99.

[51] MURPHY, P. A., and M'KAY, R. 1932. 'A Comparison of some European and American Virus Diseases of the Potato.' *Sci. Proc. Roy. Dublin Soc.*, **20**, 27, 347–58.

[52] NELSON, RAY. 1923. 'The Occurrence of Protozoa in Plants affected with Mosaic and Related Diseases.' *Mich. Agric. Exp. Sta. Tech. Bull.*, 58.

[53] —— 1932. 'Investigations in the Mosaic Disease of Beans (*Phaseolus vulgaris* L.).' *Agric. Exp. Sta. Mich. State Coll. Tech. Bull.*, 118.

[54] NOLLA, J. A. B., and ARTURO ROQUE. 1934. 'A Variety of Tobacco Resistant to Ordinary Tobacco Mosaic.' *Journ. Dept. Agric. Puerto Rico*, **17**, 4, 303.

[55] OGILVIE, L. 1928. 'A Transmissible Virus Disease of the Easter Lily.' *Ann. Appl. Biol.*, **15**, 4, 540–62.

[56] PERDRAU, J. R., and TODD, C. 1933. 'The Photodynamic Action of Methylene Blue on Certain Viruses.' *Proc. Roy. Soc. B.*, **112**, 277–98.

[57] PETERSON, ALVAH. 1934. 'A Manual of Entomological Equipment and Methods, Pt. I.' *Edwards Bros. Inc.* Ann. Arbor, Michigan.

[58] PRICE, W. C. 1932. 'Acquired Immunity to Ringspot in *Nicotiana*.' *Contrib. Boyce Thomp. Instit.*, **4**, 3, 359–403.

[59] —— 1934. 'Isolation and Study of some Yellow Strains of Cucumber Mosaic.' *Phytopath.*, **24**, 7, 743–61.

[60] RAO, M. G. V., and GOPALAIYENGAR, K. 1934. 'Studies in Spike Disease of Sandal.' *Mysore Sandal Spike Investigation Committee Bull.*, 5.

[61] REDDICK, D. 1931. 'La Transmission du Virus de la Mosaique du Haricot par le pollen.' *Deuxième Congrès Internat. de Path. Comp. Paris*, I, 363–6.

[62] RIVERS, T. M. 1932. 'Viruses.' *Science*, **75**, 1956, 654–6.

[63] —— 1932. 'The Nature of Viruses.' *Physiol. Rev.*, **12**, 3, 423–52.

[64] SALAMAN, R. N. 1933. 'Protective Inoculation against a Plant Virus.' *Nature* **131**, 468.

[65] SAMUEL, G. 1934. 'The Movement of Tobacco Mosaic Virus within the Plant.' *Ann. Appl. Biol.*, **21**, 1, 90–111.

[66] —— and BALD, J. G. 1933. 'On the Use of the Primary Lesions in Quantitative Work with Two Plant Viruses.' *Ann. Appl. Biol.*, **20**, 1, 70–99.

[67] SCHULTZ, E. S. *et al.* 1934. 'Resistance of Potato to Mosaic and other Virus Diseases.' *Phytopath.*, **24**, 2, 116–32.

[68] SEVERIN, H. H. P., and FREITAG, J. H. 1933. 'Some Properties of the Curly-top Virus.' *Hilgardia*, **8**, 1, 1–48.

[69] SHEFFIELD, F. M. L. 1933. 'The Development of Assimilatory Tissue in Solanaceous Hosts affected with Aucuba Mosaic of Tomato.' *Ann. Appl. Biol.*, **20**, 1, 57–69.

[70] SMITH, J. H. 1933. 'Some Aspects of Virus Disease in Plants.' *Empire Journ. Exp. Agric.*, **1**, 3, 206–14.

[71] SMITH, KENNETH M. 1929. 'Insect Transmission of Potato Leaf-roll.' *Ann. Appl. Biol.*, **16**, 2.

[72] —— 1931. 'Virus Diseases of Plants and their Relationship with Insect Vectors.' *Biol. Rev.*, **6**, 3, 302–44.

[73] —— 1931. 'On the Composite Nature of Certain Potato Virus Diseases of the Mosaic Group.' *Proc. Roy. Soc. B.*, **109**, 251–67.

[74] —— 1935. 'Two Strains of Streak: a Virus affecting the Tomato Plant.' *Parasitology.* (In the press.)

[75] —— and BALD, J. G. 1935. 'A Necrotic Virus Disease affecting Tobacco and Other Plants.' *Parasitology*, **27**, 2, 1–15.

[76] SPOONER, E. T. C., and BAWDEN, F. C. 1935. 'Experiments on the Serological Reactions of Potato Virus X.' *Brit. J. Exp. Path.*, **16**, 3.

[77] SREENIVASAYA, M. 1930. 'Contributions to the Study of Spike Disease of Sandal. XI. New Methods of Disease Transmission.' *Journ. Indian Instit. Sci.*, **13A**, 10, 113–17.

[78] STANLEY, W. M. 1934. 'Chemical Studies on the Virus of Tobacco Mosaic. I. Some Effects of Trypsin.' *Phytopath.*, **24**, 10, 1055.

[79] —— 1934. 'Chemical Studies on the Virus of Tobacco Mosaic. II. The Proteolytic Action of Pepsin.' *Phytopath.*, **24**, 11, 1269.

[80] STOREY, H. H. 1933. 'Investigation of the Mechanism of the Transmission of Plant Viruses by Insects. I.' *Proc. Roy. Soc. B.*, **113**.

[81] —— 1934. 'The Photodynamic Action of Methylene Blue on the Virus of a Plant Disease.' *Ann. Appl. Biol.*, **21**, 4, 588.

[82] THUNG, T. H. 1931. *Handel. 6de Nederl-Inde. Naturwetensch. Congr.*, 450–63.

[83] —— 1932. 'Analysis of the Different Epidemiology of Mosaic and Kroepoek' (in Dutch). *Proefst. voor Vorstenl. Tabak Med.*, **72**, 54 pp.

[84] TOMPKINS, C. M., GARDNER, M. W., and WHIPPLE, O. C. 1935. 'Spotted Wilt of Truck Crops and Ornamental Plants.' *Phytopath.*, **25**, 1, 17.

[85] UPPAL, B. N. 1934. 'The Movement of Tobacco Mosaic Virus in Leaves of *Nicotiana sylvestris*.' *Indian Journ. Agri. Sci.*, **4**, 5, 865.

[86] —— 1934. 'The Effect of Dilution on the Thermal Death-rate of Tobacco Mosaic Virus.' *Indian Journ. Agric. Sci.*, **6**, 5, 874.

[87] VINSON, C. G., and PETRE, A. W. 1929. 'Mosaic Disease of Tobacco.' *Bot. Gaz.*, **87**, 1, 14–38.

[88] WEBB, R. W. 1927. 'Soil Factors influencing the Development of the Mosaic Disease in Winter Wheat.' *Journ. Agric. Res.*, **35**, and *ibid.*, **36**, 1928.

[89] WHITE, P. R. 1934. 'Unlimited Growth of Excised Tomato Root Tips in a Liquid Medium.' *Plant Physiol.*, **9**.

[90] —— 1934. 'Multiplication of the Viruses of Tobacco and Aucuba Mosaics in Growing Excised Tomato Root Tips.' *Phytopath.*, **24**, 9, 1003–1011.

[91] WHITEHEAD, T., CURRIE, J. F., and MALDWYN-DAVIES, W. 1932. 'Virus Diseases in Relation to Commercial Seed Potato Production.' *Ann. Appl. Biol.*, **19**, 4, 529–49.

[92] WIGGLESWORTH, V. B. 1934. *Insect Physiology*. Methuen's Monographs on Biological Subjects.

[93] YOUDEN, W. J., and BEALE, HELEN PURDY. 1934. 'A Statistical Study of the Local Lesion Method for Estimating Tobacco Mosaic Virus.' *Contrib. Boyce Thomp. Instit.*, **6**, 3, 437–54.

[94] ZINSSER, H., and SEASTONE, C. V. 1930. 'The Influence of Oxidation and Reduction on the Virulence of Herpes Filtrates.' *Journ. Immun.*, **18**, 1, 1–9.

INDEX

Printed in Great Britain by
Butler & Tanner Ltd.
Frome and London

METHUEN'S
Books on
BIOLOGICAL AND GEOLOGICAL SUBJECTS

A SELECTION OF BOOKS ON BIOLOGICAL AND GEOLOGICAL SUBJECTS

BIOLOGY

GENERAL BIOLOGY

THE TEACHING OF BIOLOGY: A Handbook for Teachers of Junior Classes

By ETHEL M. POULTON, D.ès.Sc., M.Sc. With 106 Illustrations. *Crown 8vo.* 6s. 6d.

PLANT AND ANIMAL LIFE: An Introduction to the Study of Biology

By R. F. SHOVE, M.A. With 136 Illustrations. *Crown 8vo.* 5s. 6d.

THE STUDY OF NATURE WITH CHILDREN

By M. G. CARTER, B.Sc. *Crown 8vo.* 3s. 6d.

OUTLINES OF BIOLOGY

By Sir PETER CHALMERS MITCHELL, LL.D., F.R.S., Secretary of the Zoological Society. Revised and Supplemented by G. P. MUDGE, A.R.C.Sc. With 11 Plates and 74 Diagrams. *Crown 8vo.* 7s. 6d.

THE STUDY OF LIVING THINGS: Prolegomena to a Functional Biology

By E. S. RUSSELL, M.A., D.Sc. *Crown 8vo.* 5s. *net.*

THE MECHANISM AND PHYSIOLOGY OF SEX DETERMINATION

By RICHARD GOLDSCHMIDT. Translated by W. J. DAKIN, D.Sc. With 113 Illustrations. *Royal 8vo.* 21s. *net.*

PROBLEMS OF RELATIVE GROWTH

By JULIAN S. HUXLEY, M.A., Honorary Lecturer in Experimental Zoology, King's College, London. With 105 Illustrations. *Demy 8vo.* 12s. 6d. *net.*

WHAT IS MAN?

By Sir J. ARTHUR THOMSON, M.A., LL.D. *Crown 8vo.* 6s. 6d. *net.*

THE OPPOSITE SEXES: A Biological and Psychological Study

By ADOLF HEILBORN. With 18 Illustrations. *Crown 8vo.* 6s. *net.*

BIOLOGICAL CHEMISTRY: The Application of Chemistry to Biological Problems

By H. E. ROAF, M.D., D.Sc., Professor of Physiology, London Hospital Medical College. With 47 Diagrams. *Crown 8vo. 10s. 6d. net.*

MODERN SCIENCE: A General Introduction

By Sir J. ARTHUR THOMSON. With 6 Plates and 29 Illustrations in the Text. *Crown 8vo. 6s. net. School Edition, 3s. 6d.*

THE GREAT BIOLOGISTS

By Sir J. ARTHUR THOMSON. *Fcap. 8vo. 3s. 6d. net. School Edition, 2s. 6d.*

SCIENCE FROM AN EASY CHAIR: First Series.

By Sir RAY LANKESTER, K.C.B., F.R.S. With 87 Illustrations. *Crown 8vo. 7s. 6d. net. Cheap Edition, 2s. 6d. net.*

SCIENCE FROM AN EASY CHAIR: Second Series.

By Sir RAY LANKESTER. With 55 Illustrations. *Crown 8vo. 7s. 6d. net.* Also as 'More Science from an Easy Chair.' *2s. net.*

SOME DIVERSIONS OF A NATURALIST

By Sir RAY LANKESTER. With a Frontispiece in Colour and 21 other Illustrations. *Crown 8vo. 2s. 6d. net.*

SECRETS OF EARTH AND SEA

By Sir RAY LANKESTER. With numerous Illustrations. *Crown 8vo. 8s. 6d. net.*

GREAT AND SMALL THINGS

By Sir RAY LANKESTER. With 38 Illustrations. *Crown 8vo. 7s. 6d. net.*

RANDOM GLEANINGS FROM NATURE'S FIELDS

By W. P. PYCRAFT. With 90 Illustrations. *Crown 8vo. 7s. 6d. net.*

MORE GLEANINGS FROM NATURE'S FIELDS

By W. P. PYCRAFT. With 100 Illustrations. *Crown 8vo. 7s. 6d. net.*

METHUEN'S MONOGRAPHS ON BIOLOGICAL SUBJECTS

Fcap. 8vo. 3s. 6d. net each

General Editor: G. R. DE BEER, M.A., D.Sc., Fellow of Merton College, Oxford.

SOCIAL BEHAVIOUR IN INSECTS

By A. D. IMMS, M.A., D.Sc., F.R.S.

MENDELISM AND EVOLUTION

By E. B. FORD, M.A., B.Sc.

MICROBES AND ULTRAMICROBES
By A. D. Gardner, M.A., D.M.

THE BIOCHEMISTRY OF MUSCLE
By D. M. Needham, M.A., Ph.D. (5s. *net.*)

RESPIRATION IN PLANTS
By W. Stiles, M.A., Sc.D., F.R.S., and W. Leach, D.Sc.

SEX DETERMINATION
By F. A. E. Crew, M.D., D.Sc.

THE SENSES OF INSECTS
By H. Eltringham, M.A., D.Sc., F.R.S.

PLANT ECOLOGY
By W. Leach, D.Sc.

CYTOLOGICAL TECHNIQUE
By J. R. Baker, M.A., D.Phil.

MIMICRY AND ITS GENETIC ASPECT
By G. D. Hale Carpenter, D.M., and E. B. Ford, M.A., B.Sc.

THE ECOLOGY OF ANIMALS
By Charles Elton, M.A.

CELLULAR RESPIRATION
By N. U. Meldrum, M.A., Ph.D.

PLANT CHIMAERAS AND GRAFT HYBRIDS
By W. Neilson Jones, M.A.

INSECT PHYSIOLOGY
By V. B. Wigglesworth, M.A., M.D.

In Preparation

BIRD MIGRATION
By W. B. Alexander, M.A.

MYCORRHIZA
By J. Ramsbottom, O.B.E.

TISSUE CULTURE
By E. N. Willmer, M.A.

ZOOLOGY

ELEMENTARY ZOOLOGY
By Oswald H. Latter, M.A. With 114 Diagrams. *Demy* 8*vo.* 12*s.*
Also in Two Parts: I. Introduction to Mammalian Physiology, 4*s.* 6*d.* II. Introduction to Zoology, 8*s.* 6*d.*

SOME MINUTE ANIMAL PARASITES OR UNSEEN FOES IN THE ANIMAL WORLD

By H. B. Fantham, D.Sc., B.A., A.R.C.S., F.Z.S., Liverpool School of Tropical Medicine; and Annie Porter, D.Sc., F.L.S., Quick Laboratory, Cambridge. With 57 Diagrams. *Crown 8vo.* *7s. 6d. net.*

THE SNAKES OF EUROPE

By G. A. Boulenger, LL.D., D.Sc., Ph.D., F.R.S. With 14 Plates and 42 Diagrams. *Crown 8vo.* *7s. 6d. net.*

ENTOMOLOGY

INSECT TRANSFORMATION

By George H. Carpenter, D.Sc., M.R.I.A. With 4 Plates and 124 other Illustrations. *Demy 8vo.* *12s. 6d. net.*

A GENERAL TEXT-BOOK OF ENTOMOLOGY

Including the Anatomy, Physiology, Development, and Classification of Insects.

By A. D. Imms, M.A., D.Sc., F.R.S., University Reader in Entomology, Cambridge. With 624 Illustrations. *Royal 8vo.* *36s. net.*

BOTANY

BRITISH PLANTS: Their Biology and Ecology

By J. F. Bevis, B.A., B.Sc., and H. J. Jeffery, A.R.C.Sc., F.L.S. With 115 Illustrations. *Demy 8vo.* *7s. 6d.*

A TEXT-BOOK OF PLANT BIOLOGY

By W. Neilson Jones, M.A., F.L.S., and M. C. Rayner, D.Sc. With 42 Diagrams. *Crown 8vo.* *7s.*

HOW TO KNOW THE FERNS

By S. Leonard Bastin. With 33 Illustrations. *Fcap. 8vo.* *2s. net.*

FUNGI AND HOW TO KNOW THEM: An Introduction to Field Mycology

By E. W. Swanton. With 16 Coloured and 32 Black and White Plates. *Crown 8vo.* *10s. 6d. net.*

PLANT PARASITIC NEMATODES, and the Diseases they Cause

By T. Goodey, D.Sc., Principal Research Assistant, Institute of Agricultural Parasitology, St. Albans. With a Foreword by R. T. Leiper, M.D., D.Sc., F.R.S. With 147 Illustrations. *21s. net.*

PHYSIOLOGY

INTERFACIAL FORCES AND PHENOMENA IN PHYSIOLOGY

By Sir WILLIAM M. BAYLISS, M.A., D.Sc., LL.D., F.R.S. With 7 Diagrams. *Crown 8vo. 7s. 6d. net.*

A MANUAL OF HISTOLOGY

By V. H. MOTTRAM, M.A., Professor of Physiology at the University of London. With 224 Diagrams. *Demy 8vo. 14s. net.*

HYGIENE

A TEXTBOOK OF HYGIENE FOR TRAINING COLLEGES

By MARGARET AVERY, B.Sc. (Lond.), M.R.San.Inst. With 100 Illustrations. *Crown 8vo.* 6*s.*

A MANUAL OF HYGIENE

By Sir WILLIAM H. HAMER, F.R.C.P., and C. W. HUTT, M.D. With 94 Illustrations. *Demy 8vo. 30s. net.*

THE SCIENCE OF HYGIENE

A Text-book of Laboratory Practice for Public Health Students. By W. C. C. PAKES, D.P.H. (Camb.), F.I.C. Revised by A. T. NANKIVELL, D.P.H. (Camb.), M.D. (State Medicine). With 80 Illustrations. *Crown 8vo. 6s. net.*

PRINCIPLES AND PRACTICE OF PREVENTIVE MEDICINE

By C. W. HUTT, M.A., M.D., and H. HYSLOP THOMSON, M.D. *Two Volumes. Royal 8vo. 70s. net.* [*In the press.*

INTERNATIONAL HYGIENE

By C. W. HUTT, M.A., M.D., D.P.H., Medical Officer of Health, Metropolitan Borough of Holborn. *Demy 8vo. 10s. 6d. net.*

HEALTH: A TEXTBOOK FOR SCHOOLS

By M. AVERY. With 79 Illustrations. *Crown 8vo. 3s. 6d.*

PALÆONTOLOGY

GLOSSARY AND NOTES ON VERTEBRATE PALÆONTOLOGY

By S. A. PELLY, M.A. *Fcap. 8vo. 5s. net.*

INVERTEBRATE PALÆONTOLOGY: An Introduction to the Study of Fossils

By HERBERT L. HAWKINS, M.Sc., F.G.S., Professor of Geology, University College, Reading. With 16 Plates. *Crown 8vo. 6s. 6d. net.*

GEOLOGY

THE SCIENTIFIC STUDY OF SCENERY

By J. E. MARR, D.Sc., F.R.S. With 23 Plates and 41 Diagrams. *Crown 8vo. 7s. 6d.*

ENGLISH COASTAL EVOLUTION

By E. M. WARD, M.A., B.Sc. With 8 Plates and 33 Maps and Plans. *Crown 8vo. 8s. 6d. net.*

THE ORIGIN OF THE CONTINENTS AND OCEANS

By ALFRED WEGENER. Translated by J. G. A. SKERL, M.Sc. With 44 Illustrations. *Demy 8vo. 10s. 6d. net.*

METAMORPHISM : A Study of the Transformation of Rock-Masses

By ALFRED HARKER, M.A., F.R.S., LL.D. With 185 Diagrams. *Demy 8vo. 17s. 6d. net.*

METHUEN'S GEOLOGICAL SERIES

General Editor: J. W. GREGORY, D.Sc., F.R.S., Emeritus Professor of Geology in the University of Glasgow

THE PRINCIPLES OF PETROLOGY

By G. W. TYRRELL, A.R.C.Sc., Ph.D., F.G.S., F.R.S.E. With 78 Diagrams. *Crown 8vo. 10s. net.*

THE ELEMENTS OF ECONOMIC GEOLOGY

By J. W. GREGORY. With 63 Diagrams. *Crown 8vo. 10s. net.*

THE NAPPE THEORY IN THE ALPS : Alpine Tectonics, 1905–1928

By Professor FRANZ HERITSCH. Translated by P. G. H. BOSWELL. With 16 Illustrations and 48 Maps and Diagrams. *Crown 8vo. 14s. net.*

THE STRUCTURE OF ASIA

Contributions by F. E. SUESS, H. DE BÖCKU, D. I. MUSHKETOV, W. D. WEST, G. B. BARBER, C. P. BERKEY, and H. A. BROUWER. Edited by J. W. GREGORY. With 8 Illustrations, 18 Folding Maps and 18 Diagrams. *Crown 8vo. 15s. net.*

GENERAL STRATIGRAPHY

By J. W. GREGORY and B. H. BARRETT, M.A., B.Sc. With 36 Illustrations and 10 Maps. *Crown 8vo. 10s. net.*

DALRADIAN GEOLOGY : The Dalradian Rocks of Scotland and their Equivalents in other Countries

By J. W. GREGORY. With 20 Illustrations and 2 Maps. *Crown 8vo. 12s. 6d. net.*

THE UNSTABLE EARTH : Some Recent Views in Geomorphology

By J. A. STEERS, M.A. With 66 Maps and Diagrams. *Crown 8vo. 15s. net.*

This catalogue contains only a selection of our Biological and Geological Works. A complete list can be obtained on application to MESSRS. METHUEN & CO., LTD., 36 ESSEX STREET, LONDON, W.C.2

1234